# SUMS Readings

SUMS Readings is a collection of books that provides students with opportunities to deepen understanding and broaden horizons. Aimed mainly at undergraduates, the series is intended for books that do not fit the classical textbook format, from leisurely-yet-rigorous introductions to topics of wide interest, to presentations of specialised topics that are not commonly taught. Its books may be read in parallel with undergraduate studies, as supplementary reading for specific courses, background reading for undergraduate projects, or out of sheer intellectual curiosity. The emphasis of the series is on novelty, accessibility and clarity of exposition, as well as self-study with easy-to-follow examples and solved exercises.

Dirk van Dalen · Mark van Atten · Craig Smoryński

# Intuitionistic Analysis

## A Constructive Frame of Mind

 Springer

Dirk van Dalen
Department of Philosophy
University of Utrecht
Utrecht, The Netherlands

Mark van Atten
Archives Husserl (CNRS/ENS)
CNRS
Paris, France

Craig Smoryński
Westmont, IL, USA

ISSN 1615-2085          ISSN 2197-4144  (electronic)
Springer Undergraduate Mathematics Series
ISSN 2730-5813          ISSN 2730-5821  (electronic)
SUMS Readings
ISBN 978-3-032-16490-2          ISBN 978-3-032-16491-9  (eBook)
https://doi.org/10.1007/978-3-032-16491-9

Mathematics Subject Classification: 03-01, 03F55

Translation and revision of the Dutch language edition: "Intuïtionistische Analyse" by Dirk van Dalen, © Epsilon Uitgaven 2011. Published by Epsilon Uitgaven. All Rights Reserved.



This Springer imprint is published by the registered company Springer Nature Switzerland AG
The registered company address is: Gewerbestrasse 11, 6330 Cham, Switzerland

If disposing of this product, please recycle the paper.

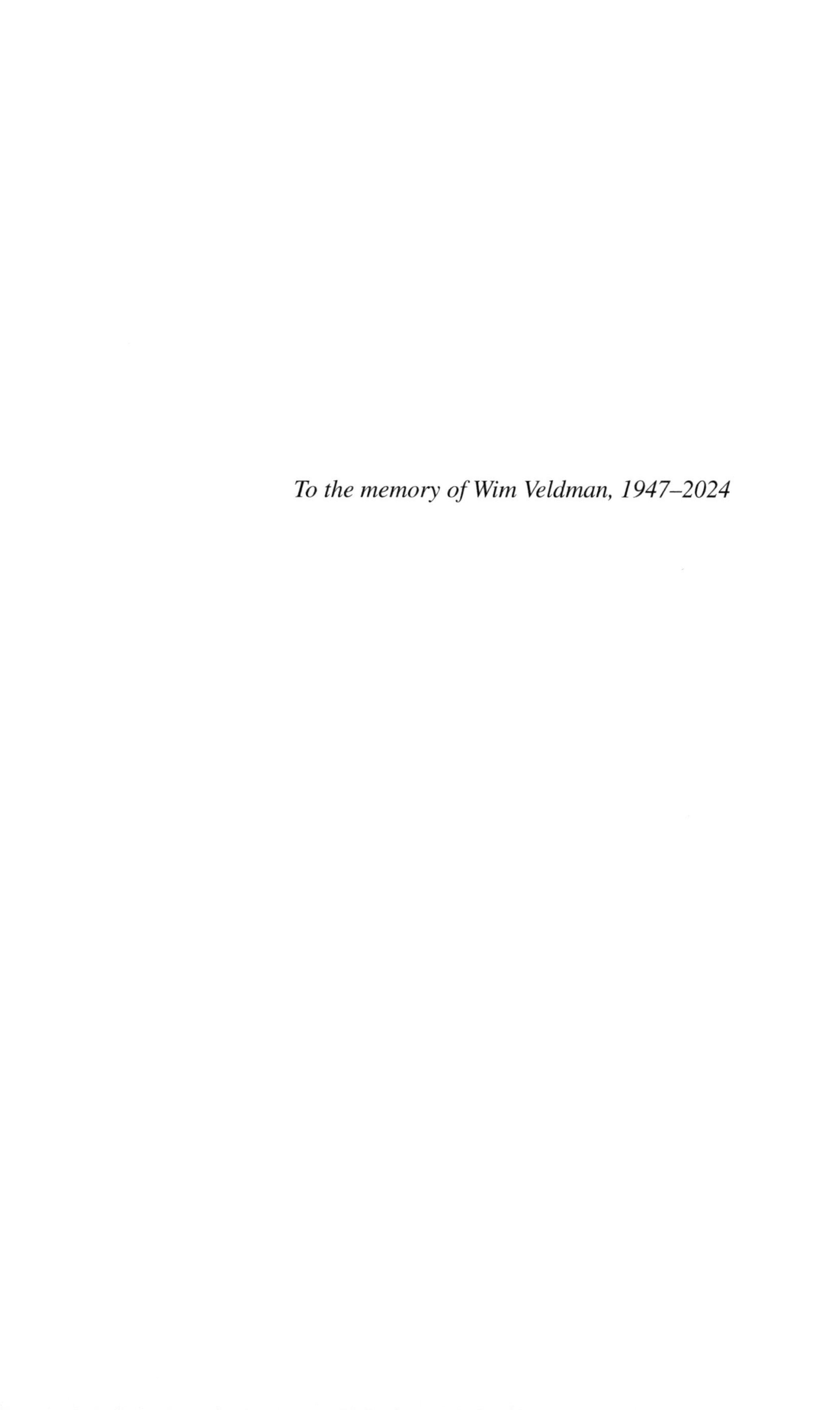

*To the memory of Wim Veldman, 1947–2024*

# Preface

We aim to acquaint the reader with fundamental ideas, questions, and mores of constructive mathematics, and of Brouwer's intuitionistic analysis in particular. Whence the choice of intuitionistic analysis as both our inspiration and ultimate theme, and not just the part it has in common with other varieties of constructivism? That is on historical and philosophical grounds. Intuitionism is the oldest systematic constructive approach to logic and mathematics, and its tradition has been offering a great number of fundamental questions and considerations that inquiring minds find too tempting to avoid. Brouwer's philosophical foundation is, whatever else one thinks of it, rich and coherent, and we find his construction of the mathematical universe based on it to be not inferior in strength and beauty to any theory presented so far.

This book is about that mathematics, not about those grounds. However, they have influenced the content and structure, on which we shall now make some remarks.

What we hope to offer is an elementary introduction to intuitionism's mathematical side that is philosophically aware. We find it of the essence to describe Brouwer's mathematics together with some of the philosophical ideas that animate it; one may accept or reject Brouwer's mathematics, but either way his project will not be understood properly if seen in isolation from the philosophical basis he gave it. In turn, one may accept or reject that basis, but in Brouwer's hands it led to the introduction of the first comprehensive alternative to classical mathematics, a remarkable event in intellectual history. However, we will not go in much detail about these philosophical ideas, let alone engage in an analysis. Our aspiration here is limited to giving a first impression of their content and of their importance to Brouwer's undertaking. Readers who should like to dig deeper are referred to the literature; some suggestions are found at the end of this book. Another expression of the philosophical awareness we hope to have brought to the text is that the topics of formal systems and formal semantics for intuitionistic logic are not considered as central, and are presented in an appendix.

Of the five chapters after the Introduction, the first four deal with mathematics that intuitionism in fact has in common with most other forms of constructivism. It could be argued that those chapters therefore also show that much constructive mathematics can be developed without having to bring in some of Brouwer's controversial innovations such as choice sequences and Creating Subject arguments. Of course, we do not disagree with that. But our goal is not to avoid these innovations, but to offer an introduction to them. And we think that, as a matter of pedagogy, their introduction benefits from an introduction to the common part.

It is also here that a main historical consideration comes in. The systematic development of constructive mathematics and of what is nowadays known as the common part began with intuitionism, and the other, later forms of constructivism followed suit. And within intuitionism itself, the development began with that part. So our first four chapters correspond to the historically first stage of intuitionism, in which a systematic criticism of classical logic and mathematics was leveraged and initiated a partial reconstruction of them; and the fifth chapter corresponds to the second stage, in which on positive philosophical grounds objects and principles were introduced that had neither been known in classical mathematics, nor could have.

Our intended audience is readers who, coming from mathematics or philosophy, are interested in the area where the two meet. Mathematically, most of this book is accessible to early university students. Since logic is not a standard part of the training everywhere, we offer a short introduction to the subject. By the above, no philosophical background is presupposed, but for those who have it, we here name Kant and Husserl as in some ways kindred spirits. Overall, the level of difficulty of the text varies with the themes, and the reader can safely leave various details for later.

This book has its origin in a Dutch version by one of us [99]. To make that accessible to a larger audience, we have translated it into English, and also taken the occasion to revise its content. The changes mostly consist in some reorganization, shifts of emphasis, additions, and small corrections. In making these revisions, we have taken into account two useful reviews of the Dutch edition, [78, 106], and benefited from remarks by Wim Veldman and Wu Jinyue. Ben Vagle provided helpful software and comments for the preparation of the index.

We are also grateful to the publisher of the Dutch edition, Epsilon Uitgaven (Utrecht), for permission to publish this new version in English, and to our editor at Springer, Remi Lodh, for his encouragement and patience.

It was a pleasure to receive the comments, corrections, and criticisms of our manuscript from meticulous and generous readers for Springer. We are much obliged to them. Any remaining sin of commission or omission is ours.

Before signing off, we wish to share a hitherto unpublished passage, a true manifesto, from Brouwer's notes for a speech at the graduation of his student Wilfrid Wilson, a hundred years ago. We present it not so much to invite the reader to pick a side in the polemic it contains, but because it illustrates so well Brouwer's drivenness in developing intuitionistic mathematics, and how he saw its place in the greater scheme of things:

> In your fatherland, where topology had no practitioners yet, you developed independently, without any external encouragement, an interest in this young branch of the mathematical sciences. This fact alone points to an ingenuous honesty of thought, and to a longing for deeper mental satisfaction than the study of classical mathematics can give. For the latter study turns

out to be incapable of delivering to the uninhibited, introverted thinker anything but norms of mechanization by humans of nature and of their fellow human beings, without further justification than instincts that lead to it, and the resulting population growth and uglification of nature. Stripped of its applications and the thought of them, classical mathematics is a meaningless building of words that induces neither mental security nor mental experience nor mental constructions. The role of classical mathematics in a country is therefore not that of a science of the mind, but of a social science, namely a regulator of social struggle and social order. In contrast, topology and intuitionism, which accompany independently of language and of practical applications a self-development of pure thought, belong to the sciences of the mind or to the philosophical sciences, to which therefore any mathematician whose ideals are not social ones, and whose flight of ideas is not hindered by ambition, feels irresistibly attracted [24, translation ours].

Vleuten, The Netherlands<br>
Saint-Germain-en-Laye, France<br>
Westmont, IL, USA<br>
December 2025

Dirk van Dalen<br>
Mark van Atten<br>
Craig Smoryński

**Competing Interests**  The authors have no competing interests to declare that are relevant to the content of this manuscript.

# Contents


**1 Introduction** ......................................................... 1

**2 Logic** ............................................................... 7
   2.1   Classical Logic ............................................. 8
   2.2   And Now Intuitionistic Logic ............................... 14
   2.3   Intuitionistic Logic is Not Three-Valued ................... 19
   2.4   Indirect Proof ............................................. 21
   2.5   A Concluding Remark ........................................ 23

**3 Natural Numbers, Integers, and Rationals** ........................... 25
   3.1   Natural Numbers ............................................ 25
   3.2   Integers and Rational Numbers .............................. 34
   3.3   More on Orderings .......................................... 36

**4 Real Numbers** ...................................................... 39
   4.1   Going Beyond the Rational Numbers .......................... 39
   4.2   The Axiom of Countable Choice .............................. 40
   4.3   Cauchy Sequences ........................................... 42
   4.4   From Cauchy Sequences to Real Numbers ...................... 43
   4.5   Ordering the Real Numbers .................................. 45
   4.6   Operations and the Apartness Relation ...................... 51
   4.7   Limits ..................................................... 58
   4.8   Supremum and Infimum ....................................... 63
   4.9   Higher Dimensions .......................................... 69

**5 Functions and Continuity** .......................................... 71
   5.1   The Concept of Continuity .................................. 71
   5.2   Continuous Functions ....................................... 74
   5.3   Discontinuity .............................................. 81
   5.4   Uniform Continuity ......................................... 82
   5.5   Differentiability .......................................... 86

5.6    Brouwer's Fixed Point Theorem ............................... 91
       5.6.1   The Classical Theorem (One-Dimensional) ............. 91
       5.6.2   A Weak Counterexample .............................. 94
       5.6.3   The Constructive Theorem (One-Dimensional) .......... 94
       5.6.4   The Constructive Theorem (Two-Dimensional) .......... 98

**6   Going Forth** ................................................. 105
   6.1    The Basis of Brouwer's Intuitionistic Mathematics: The
       Two Acts ................................................. 105
   6.2    Spreads and Fans ......................................... 109
   6.3    The Continuity Principle .................................. 110
   6.4    The Fan Theorem and Some Applications ..................... 113
   6.5    The Brouwer-Kripke Schema ............................... 120
   6.6    More About Discontinuity ................................. 124
   6.7    More About the Unsplittability of the Continuum .............. 126
   6.8    More About the Axiom of Choice ........................... 128
   6.9    Intuitionism Compared with Two Other Currents
       in Constructive Mathematics ............................... 130
       6.9.1   Recursive Mathematics and Russian Constructivity ...... 130
       6.9.2   Bishop's School ................................... 136
   6.10   Closing Remark ......................................... 137

**Appendix A: Two Technical Devices in the Study of
                Intuitionistic Logic** .................................. 139

**References and Further Reading** ................................. 153

**Index** ......................................................... 159

# Chapter 1
# Introduction

**Abstract** This chapter introduces constructive mathematics and intuitionism, the foundational program of L.E.J. Brouwer. Intuitionism views mathematics as a mental activity of construction, independent of language or logic. We contrast this with classical mathematics, whose reliance on logic leads to non-effective existence proofs. Through "weak counterexamples", based for example on undecided questions regarding the decimal expansion of Pi, we show that classical principles like the law of trichotomy are not constructively valid. This critique sets the stage for the development of constructive and intuitionistic analysis.

Constructive mathematics, as an independent discipline, is not that old yet. Of course, all kinds of algorithms had found their place in the treasure chest of traditional mathematics, and practitioners have always taken pride in actually being able to perform certain calculations or constructions. But the question of a systematic practice of constructive, or effective, mathematics only arose when abstract methods were introduced on a large scale to avoid laborious calculations. The new methods that appeared in the second half of the nineteenth century brought ever greater abstraction, and ever more powerful tools for proving the existence of certain mathematical objects. One has to think of a rich variety of notions—groups, invariants, fixed points, spaces, cardinal numbers, etc. Only when it became clear how far mathematics had moved away from the concrete and constructible did a feeling of unease arise.

By the late 19th century, the German Leopold Kronecker was the first to insist that mathematics should be effective throughout, on philosophical grounds: "God made the integers, all the rest is the work of man!" In more recent times, the advent of computer science and programs for computers have made that unease very palpable–the difference between effective and non-effective turned out to be highly significant.

But, in between and above all, the paradoxes found at the beginning of the twentieth century testified to the precarious situation that mathematics had grown into. In France, Henri Poincaré, Jules Molk, Émile Borel and others thought about putting limitations on overly abstract mathematics.

© The Author(s), under exclusive license to Springer Nature Switzerland AG 2026
D. van Dalen et al., *Intuitionistic Analysis*, Springer Undergraduate Mathematics Series,
https://doi.org/10.1007/978-3-032-16491-9_1

However, the Dutchman L. E. J. Brouwer[1] was the first to develop a comprehensive system of constructive mathematics from the ground up, backed up by philosophical views that were coherent and deep enough to make such a development possible.

In his dissertation "On the Foundations of Mathematics" from 1907 he put forward a number of ideas, only two of which we will mention here:

(1)  Mathematics consists first of all in acts of construction in the human mind.
(2)  The unlimited faith in logic has led to serious weakening within mathematics.

The first point is *positive*; it is about the nature of mathematical acts and the objects resulting from them. The second is *negative*; it points out imperfections in the modern mathematical practice of its day.

The first point is the basis of Brouwer's *intuitionistic mathematics*. He took the position that mathematics is, not in practice but in its essence, an introspective affair, consisting in making the objects and their properties in one's mind. The building material is given by what the mind sees in itself, and it is ultimately(!) also by seeing that the mind comes to know what works and what does not. In philosophical jargon, this seeing is a form of intuition, and this explains the name of Brouwer's foundational program, *intuitionism*. Brouwer was not content with collecting clever methods to construct certain mathematical objects or to solve certain problems, but offered a general framework within which all of constructive mathematics could be realized.

The second point is important for much of what is covered in this book. It has to do with the *effectiveness* or *constructivity* of mathematics. This is most easily demonstrated by the concept of *existence*. When we say "the equation $\sin(x) = 0.7231$ has a solution", we mean "there is a real number $a$ such that $\sin(a) = 0.7231$", and we can calculate that number too. We can point to that number $a$, that is to say, it can be approximated arbitrarily well. "There is", or "there exists", imposes an obligation, namely to produce the intended object. For a long time mathematics has taken a somewhat more liberal stance, that of "there is an $x$ satisfying $f(x) = 0$" just means "it is not true that, for all $x$, $f(x) \neq 0$". That seems to be the same; after all isn't "not for all not" the same as "there is"? Those who can be content with that need not worry immediately, but they are cheating. They say there is something there, but they can't show it and cannot even promise that they will find it. That is a denaturing of the concept of existence; whoever claims there is a Martian living in his garden shed must also be able to show it. Then why is mathematics so gullible? The answer seems to us, "because logic approves".

When Brouwer started his crusade against ineffective methods, computers did not yet exist. Algorithms had been around for a long time–they are as old as mathematics–but, just at the end of the nineteenth century, general abstract methods, which tend to circumvent algorithms, came into vogue. Brouwer did not enter the arena to fight paradoxes, but to lead mathematics back to its origin: human thinking. His position

---

[1] Here we write, as is often seen in the literature, his initials instead of his first names. They were Luitzen Egbertus Jan.

was not an easy one–he had to promote reform at a time of great mathematical prosperity. Algebra, analysis, geometry, and set theory flourished as never before. To begin with, he had to show that there were indeed problems even in ordinary mathematics. For this purpose he invented a tool which is now known as the "weak counterexample". This is a convenient way to show that some mathematical problems did not (yet) have a solution because otherwise a genuine, objective mathematical question, to which we had as yet no answer, would be affirmed or refuted. His argument used "sequences in $\pi$" (but others are possible).

The number $\pi$ is notoriously mysterious. It is the ratio of the perimeter of the circle to the diameter. Little or nothing is known about regularities in its decimal expansion. Decimals of $\pi$ were calculated early on and, with the advent of the computer, the list of decimals quickly became longer. Now one can look at those decimals and ask all kinds of questions: Is there a digit that occurs only finitely often? Do two of the same digits appear one after the other infinitely often? Does a sequence 0123456789 occur? Do 20 nines occur in succession? Are some numbers eventually more common than others? Etc. All those questions were unanswered in 1908 when Brouwer started looking for examples. He often used the question about the presence of 0123456789 in $\pi$, but that is no longer possible: in 1997 a computer programmed by Takahashi and Kanada found out that it *does* occur, starting at the 17,387,594,880th digit after the decimal point; see [12]. That does not change the principle, of course; we have been using other examples ever since, and they abound.

Here is a typical weak counterexample , which we will make use of several times. Calculate the decimals of $\pi$ one after the other, and in parallel those of a real number $a$, constructed as follows. Every time a new decimal $x$ of $\pi$ is found, we choose a new decimal $y$ of $a$: if $x$ is the first decimal preceded by 20 consecutive nines, we take $y = 1$, and otherwise $y = 0$. So if and when we stumble upon 20 consecutive nines, we get, perhaps[2]:

$$\pi = 3.14159265358979323\ldots\overbrace{999\ldots997}^{20\times}\ldots$$
$$a = 0.000\ldots\ldots\ldots\ldots\ldots\ldots\ldots\ldots0010..$$

But at present we do not know whether this will happen. What we do know is that this number $a$ is very small: it starts out with many zeros. We would now like to know if $a > 0$. That clearly depends on $\pi$, and we immediately see:

$a > 0 \leftrightarrow$ there is a sequence of 20 nines in the decimal expansion of $\pi$.

Since it is unknown whether such a sequence occurs, it is unknown whether $a > 0$.

*We may therefore not claim that $a$ is positive.*

---

[2] "Perhaps"–because the trailing 7 is just a guess.

What about $a = 0$? In that case all decimals of $a$ must be 0, but to know that, we must know that there is no sequence of 20 nines in $\pi$ (i.e., in the decimal expansion of $\pi$). Unfortunately, that, too, is still unknown. So:

*We also may not claim that $a$ equals 0.*

But what do we know about $a$? In any case, that it is not smaller than 0 (why?). Let us now ask ourselves if $a$ is rational. So far, we don't know whether $a = 10^{-n}$ for some $n$ (recall that by definition of $x$, there can be at most one 1 in $a$), nor whether $a = 0$; so we can't give the answer (yet). But might $a$ be irrational? Then $a$ may not be one of the numbers $10^{-n}$, in which case there is no sequence of 20 nines in $\pi$, but then $a$ is equal to 0. However, then $a$ is rational. Contradiction. This establishes that $a$ cannot be irrational, i.e., $a$ is *not-not-rational*. So we have learned at least *something* about $a$. With this we have a weak counterexample to the claim that a double negation amounts to an affirmation,[3] and that *Reductio Ad Absurdum* is not effective. Logic seems to be in trouble.

Let us run another example of this type, this time with two numbers $a$ and $b$. The number $a$ is defined as above. The definition of $b$ is a variation: for the $n$-th decimal of $b$ we take 0, except if the $n$-th decimal of $\pi$ is immediately preceded by a sequence of 20 eights, in which case we take 1. So if both sequences occur, and that of the nines comes first, the situation would, perhaps, be this:

$$
\begin{aligned}
\pi &= 3.14159265358979323\ldots\overbrace{999\ldots997}^{20\times}\ldots\overbrace{888\ldots880}^{20\times}385\ldots\\
a &= 0.000\ldots\ldots\ldots\ldots\ldots\ldots\ldots\ldots0010\ldots\ldots\ldots\ldots\ldots\\
b &= 0.000\ldots\ldots\ldots\ldots\ldots\ldots\ldots\ldots\ldots\ldots\ldots0010\ldots\ldots
\end{aligned}
$$

But at present, it is not known whether such sequences occur in $\pi$, and, in case both do, which one comes first.

Question: is $a > b$, $a < b$ or $a = b$? (When would we have $a = b$?)

It could happen that we do not know whether either of these sequences occurs, but somehow managed to prove, for example, that if a sequence of eights occurs at all, it can only occur after a sequence of nines. However, nothing of the kind is known. So $a > b$, $a < b$ and $a = b$ are all three unknown. Thereby, we have a weak counterexample to the mathematical law of the *trichotomy*, which claims that we always have $a > b$ or $a = b$ or $a < b$.

We are now suspiciously close to a crisis of confidence in mathematics. Do the customary laws still hold? The answer is: some common mathematical laws are indeed not constructively (or effectively) valid, and even logic itself fails!

It will now be clear that the problems of constructive mathematics are strongly related to the infinite; as long as we are dealing with finite discrete sets we can answer questions just by checking all cases. (A discrete set is one with a decidable equality relation on its elements.) But if we must look at an unlimited number of

---

[3] It does in the case of *decidable* statements, i.e., those for which we have a method that will either prove or disprove them.

cases ("unbounded search"), there is no guarantee that we can give an answer in a finite time.

In the rest of this book we will examine to what extent the example of intuitionistic arithmetic can be followed. We want to see whether common theorems of ordinary mathematics are *constructive* or *effective*. In particular, we wish to know whether existence statements–of the form, "there is an $x$ satisfying the condition $A(x)$"– can be constructively demonstrated. These kinds of questions are at the center of mathematics, and we come across them everywhere: there is a complex number $z$ such that $z^n + a_{n-1}z^{n-1} + \cdots + a_1z + a_0 = 0$ (Fundamental Theorem of Algebra); there is a common tangent to two given circles; a continuous function on a closed interval has a maximum; etc. Constructive mathematics considers its task to be to exhibit those promised things.

In the first half of the twentieth century the association of constructive mathematics with Brouwer was so strong that "intuitionistic" became a common synonym for "constructive". That is going too far, and does justice to neither set of ideas. This became clear when after Brouwer, alternative systematic proposals were put forward to constructivize mathematics, notably by the Russian mathematician Andrey Markov and the American mathematician Errett Bishop. The systems of Brouwer, Markov, and Bishop not only are underpinned by different motivations and views, and not only use different methods, but lead to different results. At the same time, there is a common part.

Constructive mathematics and intuitionistic mathematics stand in contrast to *classical mathematics*, which employs the traditional, non-effective logic (classical logic) and thereby produces non-effective mathematics. To constructivists, much of classical practice seems supernatural. Classical mathematicians believe in a fixed, wholly determinate mathematical universe, existing independently of our acts and constructions, and where each object has its properties whether we know them or not. In philosophical terms, classical mathematicians are "Platonists" or "realists". There are all sorts of reasons why one may prefer to take up a constructive position instead. Intuitionists do so because they regard mathematics as a human activity consisting in making certain mental constructions. The differences between such views–between the classical and the constructive, but also, among the latter, between the intuitionistic and the non-intuitionistic–for the mathematics and logic one gets are not small.

In this book we will introduce the reader to intuitionistic analysis. Chapter 2 through 5 deal with parts that intuitionism shares with a more general constructive approach. Specifically intuitionistic views, concepts, and theorems are discussed in Chap. 6. Depending on one's view, these are a deepening of the common part, or additions to it.

At times, we will consider things classically–for the sake of contrast, to show what the still dominant view and practice are that intuitionism, and the wider constructive developments it inaugurated, part ways with.

*A terminological note.*
In this book we will have occasion to speak of a variety of conceptions of mathematics, each with their own conception(s) of collection. Until Chap. 5, we will simply use the

term "set", and also "equivalence class", in a general constructive sense; we hardly have occasion to speak of classical sets. In Chap. 5, we will also discuss Brouwer's two intuitionistic notions of collection: the *species* and the *spread*. The latter is actually a special instance of species. Various things that Brouwer defined or proved in their terms have been adapted to the theory of collections in other varieties of constructivism.

# Chapter 2
# Logic

**Abstract** If the classical notion of knowledge-independent truth is replaced by the constructive one of possessing a construction or proof, then logic must be explained differently. On the so-called Proof Interpretation, implication for example is now explained as a method that, whenever I have a proof of the antecedent, can be used to transform it into a proof of the consequent. The Principle of the Excluded Middle is rejected, as, given an arbitrary proposition, one is not always in a position to prove it or its negation. One might suspect that this rejection implies that propositions possess a "third" truth value, yet we present a theorem by Glivenko demonstrating that intuitionistic logic is not three-valued. We also show that while proof by contradiction is useful for establishing negative facts, it is, in general, powerless to establish positive ones.

Brouwer found a meaning for logic that avoids falling into the abyss of classical logic. That meaning turned out to be adaptable to other forms of constructivism as well. In fact the terms "intuitionistic logic" and "constructive logic" have often been used interchangeably, and still are, also by us. However, there certainly are contexts where it makes good sense to give these terms distinct meanings, and then one should. For example, "constructive logic" is usually taken to correspond to a proper part of classical logic, whereas "intuitionistic logic" in the specific sense contains strong counterexamples to principles of classical logic. We will see some in Chap. 6.

It is true that Brouwer denied logic any creative power: genuine mathematical results can always be obtained by direct mathematical reasoning, and cannot depend on the use of logical principles. On that austere but impeccable view, a book on intuitionistic mathematics need not discuss logic at all. Still, being clear about logic remains useful as a reminder of the pitfalls of classical reasoning, and logic remains a tool for efficient notation and communication. For these reasons, this first chapter will be on logic.

In the Introduction we discussed the question whether there is a sequence of 20 consecutive digits 9 in the decimal expansion of $\pi$; some rough ideas were enough to illustrate that classical logic is not effective, and intuitionistic logic is. Here we will look at both in somewhat more detail, but for a thorough treatment we refer to the chapters on propositional and predicate logic in [102].



D. van Dalen et al., *Intuitionistic Analysis*, Springer Undergraduate Mathematics Series,
https://doi.org/10.1007/978-3-032-16491-9_2

## 2.1  Classical Logic

To begin with, let us briefly explain how ordinary, classical logic interprets the basic connectivesor operators. Remember that logic has its own language, which is very economical. It lends itself perfectly to scientific use, as well as to reasoning in every-day life, but that frugality ("poverty", some might say) means that it cannot do justice to the elegant emotional aspects of our colloquial language. Only a few basic operators are needed to set up the logical language, a fact learned from experience. If you experiment a bit, you will soon see that you can express a lot with them. It is a small miracle that mathematics (and much of science) can be done with so few operators—a magnanimous gift of language and thought.

Here is the list of the usual logical operators:

$$\land \quad \text{stands for } and \text{ (conjunction)}$$
$$\lor \qquad\qquad\qquad or \text{ (disjunction)}$$
$$\rightarrow \qquad\qquad\qquad if\dots\ then\ \dots \text{ (implication)}$$
$$\neg \qquad\qquad\qquad not \text{ (negation)}$$
$$\bot \qquad\qquad\qquad a\ contradiction \text{ (falsum)}$$
$$\leftrightarrow \qquad\qquad\qquad if\ and\ only\ if \text{ (bi-implication or equivalence)}$$
$$\forall \qquad\qquad\qquad for\ all\ \dots \text{ (universal quantifier)}$$
$$\exists \qquad\qquad\qquad there\ is\ a\ \dots \text{ (existential quantifier)}$$

Falsum is a bit of an odd one out; a logical operator usually takes one or more formulas into a new formula, but here is a logical operator that is formula all by itself. Analogy: a unary function yields a value given one argument; a 0-ary function yields a value given no argument. It is a constant.

Abbreviations can be introduced for compound formulas, and in fact we take $A \leftrightarrow B$ to be short for $(A \rightarrow B) \land (B \rightarrow A)$.

The logical operators take some getting used to, but just as you utter the sounds "sine ex" without hesitation when you read "$\sin(x)$", you can learn to read formulas that use the symbolic language of logic in a few minutes. Examples:

$$
\begin{array}{ll}
\text{for } A \land B & \text{read "}A \text{ and } B\text{"} \\
\quad A \lor B & \text{"}A \text{ or } B\text{"} \\
\quad A \rightarrow B & \text{"if } A \text{ then } B\text{", or "}A \text{ implies } B\text{"} \\
\quad \neg A & \text{"not-}A\text{"} \\
\quad A \leftrightarrow B & \text{"}A \text{ if and only if } B\text{"} \\
\quad A \rightarrow (B \lor C) & \text{"if } A \text{ then } B \text{ or } C\text{"} \\
\quad \forall x(x > 0 \land x \leq 2) & \text{"for all } x,\ x \text{ is positive and } x \text{ is at most 2"} \\
\quad \exists x \forall y(x > y) & \text{"there is an } x \text{ greater than all } y\text{"}
\end{array}
$$

In a precise treatment of logic, one must fix the class of formulas with mathematical rigor, as well as related notions like subformula and operations like substitution. Since this isn't a textbook of logic, we will be rather informal about such matters.

Formulas are constructed systematically using logical operators (and parentheses). The formulas that are the basic building blocks of such a structure, i.e., formulas without logical operators, are called *atomic formulas*, or *atoms* for short. What they look like depends on the field we are trying to describe. In our case we are always dealing with simple mathematical atoms, such as $x = y, 2 + 3 \geq 2 \cdot 2$. A logical formula is an expression that can be constructed in finitely many steps beginning with atoms and using logical operators.

For example, let $A$ be any atom. Consider the formula $\neg(A \vee \neg A)$, where brackets have been used to make it clear that we did not mean the disjunction of $\neg A$ and $\neg A$. The formula can be obtained by taking $A$, combining it with $\neg$ to get $\neg A$, combining $A$ with that by means of $\vee$, and then applying $\neg$ again.

Just as in ordinary arithmetic and algebra, we often do not want to write out all parentheses that a formal definition would require. Thus, we adopt rules of order. From high to low precedence: $\neg, \wedge, \vee, \rightarrow, \leftrightarrow$. This means that $A \wedge B \rightarrow C$ is read as $(A \wedge B) \rightarrow C$, not as $A \wedge (B \rightarrow C)$. Furthermore, $\wedge$ and $\vee$ associate to the left, so $A \vee B \vee C$ is read as $(A \vee B) \vee C$. In contrast, $\rightarrow$ and $\leftrightarrow$ associate to the right, so $A \rightarrow B \rightarrow C$ is read as $A \rightarrow (B \rightarrow C)$.

The starting point of classical logic is the hypothesis that every statement is true or false. We have to show a little good will: the statements have to be "constant". For example, $\pi < \sqrt{10}$ is true, but, if no number has been assigned to the variable $x$, $\pi < x$ has no truth value.

We can use a kind of calculation method with two truth values: *true* and *false*. This logic is therefore a *bivalent logic*. As good mathematicians we prefer to work with symbols rather than words. Here we use 1 and 0 for *true* and *false*. There are also authors who instead of 1 and 0 use $t$ and $f$, or $T$ and $F$.

The truth values determine the meanings of the propositional connectives, as the logical operators other than the two quantifiers $\exists$ and $\forall$ are called. This is traditionally done by means of *truth tables*.

| $\wedge$ | 0 | 1 |
|---|---|---|
| 0 | 0 | 0 |
| 1 | 0 | 1 |

| $\vee$ | 0 | 1 |
|---|---|---|
| 0 | 0 | 1 |
| 1 | 1 | 1 |

| $\rightarrow$ | 0 | 1 |
|---|---|---|
| 0 | 1 | 1 |
| 1 | 0 | 1 |

| $\leftrightarrow$ | 0 | 1 |
|---|---|---|
| 0 | 1 | 0 |
| 1 | 0 | 1 |

| $\neg$ | |
|---|---|
| 0 | 1 |
| 1 | 0 |

| $\bot$ | 0 |
|---|---|

For falsum the table is degenerate, since it is a connective with 0 arguments. In words, the tables say:

(1)  A *conjunction is true* if both parts (conjuncts) are true.
(2)  A *disjunction is true* if at least one of its parts (disjuncts) is true.
(3)  An *implication is true* if the right-hand side has no smaller truth value than the left-hand side (the truth value does not decrease).
(4)  An *equivalence is true* if the parts have the same truth value.

(5) A *negation is true* if the statement that it is applied to is false.

(6) *Falsum* is false.

In fact, these tables define functions with zeros and ones as input. A logical operator is not a function on numbers, but on formulas. However, it defines a function, for example $\wedge : \{0, 1\} \times \{0, 1\} \to \{0, 1\}$. It is of course sloppy to indicate this function with $\wedge$ again; we will allow ourselves this sloppiness. Better to have a notation that isn't quite correct than one that complicates things uncomfortably in the long run. The French speak of an "abus de langage" in such cases. Note that the truth values of the conjunction, disjunction, and negation are easy to capture in formulas:

$$x \wedge y = \min(x, y)$$
$$x \vee y = \max(x, y)$$
$$\neg x = 1 - x$$

We don't have truth tables for the quantifiers, because we don't know in advance how many elements we are talking about. And if there are infinitely many elements, we would even need an infinite table. We simply agree that:

(1) $\forall x\, A(x)$ is true if $A(a)$ is true for all elements $a$ of the domain we are talking about (*domain of discourse*), whether we know this or not.

(2) $\exists x\, A(x)$ is true if $A(a)$ is true for at least one element $a$ from the domain, whether we can indicate $a$ or not.

Note that we often write the quantifiers without including in that notation the information what domain we're talking about, because it is often clear from the context. But if we want to, we write, for example, $\forall x \in \mathbb{N}\, A(x)$.

As tempting as it may be in mathematics to give a definition and then get down to business (i.e., applying the definition), it doesn't hurt to check whether our definitions are in agreement with everyday use. For $\wedge$, $\vee$, $\neg$ there is no problem; $A \wedge B$ is true if both parts are true, otherwise $A \wedge B$ is false. For $A \vee B$ we can make a similar consideration. It is even easier with negation: "*not-A*" is only true if $A$ is false. Now consider the implication, where there is a problem. We are used to using "if $A$ then $B$" under sensible circumstances—something like "if it's Tuesday, this must be Belgium," or "if $a$ is positive, then it has a square root". In such cases, the antecedent need not be true, but if it is true, then the consequent is also true. In ordinary life, we do not bother to specify what we could say in cases where it does not seem sensible even to assume that the antecedent is true. But in classical logic, every statement should be either true or false. Clearly, it is false that $-3$ is positive; but when someone asks if $-3$ has a square root, we say "there is no implication". Now if someone asks whether "if $-3$ is positive, then $-3$ has a square root" is true, then we have to think for a while. Either we decide that $A \to B$ is not always defined (just like $1/x$ is not defined everywhere), or we come up with values for $0 \to 1$ and $0 \to 0$. Logic chose the second solution very early on, mainly because it makes the system much more coherent. So in the truth table of $\to$ we have to determine the top row in a sensible way. The choice $0 \to 0 = 1$ is obvious; we are used to the fact that something false leads to something else that is also false. For example, consider $2 = 5 \to 1 = 2$; the proof goes as follows: add 1 in $2 = 5$ on both sides to get $3 = 6$. Now divide both sides by 3, the result is $1 = 2$.

For $0 \to 1$ it is a bit more difficult. We could say we declare $A \to B$ true if $B$ is true when $A$ is true. Now if for some reason $A$ cannot be true, then we could reason as follows: Suppose that $A$ is true after all, then we only need to check whether $B$ is true. And that is right. Unfortunately, we have now used the principle we wanted to demonstrate at a higher level. We will come back to this case later, for now we accept the simplest solution: $0 \to 1 = 1$ because we say so.[1] Practice will prove us right.

With the help of the truth tables we can verify all kinds of laws of logic. Here are some examples. To determine whether a statement is true, you need to know the truth values of its parts. $A \wedge B$ is sometimes true and sometimes false, but $A \to A$ is true for all values of $A$. After all, the implication can only be false if the first member is true and the second member is false, but here the first and second members always have the same value, so the falsehood cannot occur. A statement that is always true in this way is called a *tautology*. Tautologies are important because in practice they are the guarantees of certainty.

**Theorem 2.1** *The following are tautologies (are always true):*

*(1)  $(A \wedge A) \leftrightarrow A$*
*(2)  $(\neg\neg A) \leftrightarrow A$ ("double negation is affirmation")*
*(3)  $(A \wedge \neg A) \to B$ ("Ex Contradictione Sequitur Quodlibet", "from a contradiction everything follows")*
*(4)  $\bot \to A$ ("Ex Falso Sequitur Quodlibet", "from the false everything follows")*

### *Proof*

(1) $A \wedge A$ always has the same value as $A$, so according to the truth table of the equivalence $A \wedge A \leftrightarrow A$ always has the value 1.
(2) The negation swaps the truth values 0 and 1. When this happens twice in a row, we're back to the original value. That means that left and right side always have the same value.
(3) We first notice that $A \wedge \neg A$ always has the value 0, since $A$ and $\neg A$ never have value 1 at the same time. Now we see that according to the truth table of the implication, the truth value of the whole is always 1, independent of the truth value of the second member.

---

[1] This is a familiar procedure: "Why do I have to finish my plate?"—"Because I say so." While we're on the topic, let us quote this anecdote about Georg Kreisel, as told by Henk  Barendregt [1, p. 5]. "A few days later  Kreisel and I were invited by the baroness for dinner at her castle. Impressive stairs went down to a cellar with burning torches on the walls. Some of the dishes were exotic. During the dessert Kreisel showed a side of his that was new to me. One of the children started to whine about the pudding. "Ma, I do not want the raisins in it; Ma, I want you to take them out …" It went on and on. Then Kreisel spoke very slowly to the little one, a boy about six years old: "Listen. What do you think is worse: your mother does not want to take the raisins out; or, your mother is not able to take them out?" The little boy had to think for a while. "It is worse, if she does not want to take them out", was the answer. "Well", Kreisel continued, "it is simply the case that she is not able to take them out!" The boy could do nothing else but eat his pudding."

(4)  Like the previous one.

$\square$

In the above we could see more or less immediately that the statements were tautologies. This is no longer so easy to see for more complex statements. We have a calculation trick available for that. Instead of an explanation, we give a few examples, which clearly show how and why the method works.

- $(A \to B) \to (\neg B \to \neg A)$ is a tautology. (Law of contraposition)

**Proof**

| $A$ $B$ | $A \to B$ | $\neg B$ | $\neg A$ | $\neg B \to \neg A$ | $(A \to B) \to (\neg B \to \neg A)$ |
|---|---|---|---|---|---|
| 0  0 | 1 | 1 | 1 | 1 | 1 |
| 0  1 | 1 | 0 | 1 | 1 | 1 |
| 1  0 | 0 | 1 | 0 | 0 | 1 |
| 1  1 | 1 | 0 | 0 | 1 | 1 |

$\square$

What we did here is make a table in which we put all possible values of $A$ and $B$ together, then calculate step by step the value of the entire statement. We then see that the last column contains only ones, i.e., whatever values $A$ and $B$ take, the statement is always true.

The implication above is an everyday truth, we know many examples of it: "If Arlene is Berend's daughter, then Berend is Arlene's father, so if Berend is not Arlene's father, then Arlene is not Berend's daughter."

- $\neg(A \land B) \leftrightarrow (\neg A \lor \neg B)$ is a tautology. (De Morgan's law)

**Proof**

| $A$ $B$ | $\neg A$ | $\neg B$ | $A \land B$ | $\neg(A \land B)$ | $(\neg A \lor \neg B)$ | $\neg(A \land B) \leftrightarrow (\neg A \lor \neg B)$ |
|---|---|---|---|---|---|---|
| 0  0 | 1 | 1 | 0 | 1 | 1 | 1 |
| 0  1 | 1 | 0 | 0 | 1 | 1 | 1 |
| 1  0 | 0 | 1 | 0 | 1 | 1 | 1 |
| 1  1 | 0 | 0 | 1 | 0 | 0 | 1 |

$\square$

Here is a verbal example: "$n$ and $m$ are not both even if and only if $n$ is odd or $m$ is odd".

- $(A \to B) \leftrightarrow (\neg A \lor B)$ is a tautology.

***Proof***

| $A$ | $B$ | $A \to B$ | $\neg A$ | $\neg A \lor B$ | $(A \to B) \leftrightarrow (\neg A \lor B)$ |
|---|---|---|---|---|---|
| 0 | 0 | 1 | 1 | 1 | 1 |
| 0 | 1 | 1 | 1 | 1 | 1 |
| 1 | 0 | 0 | 0 | 0 | 1 |
| 1 | 1 | 1 | 0 | 1 | 1 |

$\square$

"If it rains, the street is wet" is true if and only if "it is not raining or the street is wet".

Note that the last tautology shows that the implication can be defined from the disjunction and negation. For that version, i.e., $\neg A \lor B$, the composition of the truth table for $\to$ is immediately clear.

**Execise 2.2** Show that the following formulas are tautologies:

(1) $\neg(A \lor B) \leftrightarrow (\neg A \land \neg B)$ (De Morgan's law)
(2) $[A \lor (B \land C)] \leftrightarrow [(A \lor B) \land (A \lor C)]$
(3) $[(A \to B) \land (B \to C)] \to (A \to C)$ (transitivity of the implication)
(4) $(A \lor B) \leftrightarrow (\neg A \to B)$
(5) $A \lor \neg A$ (PEM)

That last tautology is the Principle of the Excluded Middle, abbreviated as PEM. It is also known as the *Principle of the Excluded Third*, and by the two Latin names *Principium Tertium Non Datur* and *Principium Tertii Exclusi*. In words, it asserts "A statement is true, or its negation is true"; a third case is excluded.

**Execise 2.3** Prove that $A \to B$ is a tautology if and only if the value of $A$ is less than or equal to the value of $B$, i.e., if the truth value does not decrease.

We will show two closely related examples with classical laws for the quantifiers.

**Theorem 2.4**

*(1) $\neg\forall x\, A(x) \to \exists x\, \neg A(x)$*
*(2) $\neg\forall x\, \neg A(x) \to \exists x\, A(x)$*

***Proof*** For (1), assume that $\neg\forall x\, A(x)$. Then $\forall x\, A(x)$ is not true, so not for all $a$ in the domain $A(a)$ holds, so by PEM there is an $a$ for which $\neg A(a)$ holds, so $\exists x\, \neg A(x)$.

For (2), set $B(x) = \neg A(x)$, apply (a) to $B(x)$, and convert back to get $\neg\forall x\, \neg A(x) \to \exists x\, \neg\neg A(x)$. Then, using "double negation is affirmation" in the form $\neg\neg A(x) \to A(x)$, $\exists x\, A(x)$. $\square$

At the end of the next section, there will be a weak counterexample to this classical theorem.

## 2.2  And Now Intuitionistic Logic

The principle that came under Brouwer's attack first was PEM. For many, the principle was and remains self-evident and indispensable; as the great mathematician David Hilbert put it in 1927, "Taking the principle of excluded middle from the mathematician would be the same, say, as proscribing the telescope to the astronomer or to the boxer the use of his fists" [103, p. 476].

But let us see how PEM holds up in a test case. There is the old Goldbach Conjecture stating that every even number greater than 2 is the sum of two prime numbers. The number 1 is not generally considered a prime. The simplest cases confirm this conjecture without problems: $4 = 2 + 2, 6 = 3 + 3, 8 = 5 + 3, 10 = 7 + 3, 12 = 7 + 5, 14 = 11 + 3, \ldots$ To date no counterexamples to the Goldbach Conjecture have been found, and this is no small matter: with the help of computers, the conjecture has been verified up to huge numbers. Nevertheless, this says very little, because compared to the infinity of cases to be considered, even huge numbers are very small! Until a proof is found that covers *all* cases, i.e., that shows by some global method that all even numbers $> 2$ can be split in the manner indicated, or a counterexample is found, we do not know enough to decide the conjecture. So we must conclude that in an effective sense the statement "every even number $> 2$ is the sum of two primes or there is an even number $> 2$ that is not a sum of two primes" cannot be accepted yet, because there is still no decision between the two outcomes. As long as there are such cases, there is no reason to accept PEM.

Now there are mathematicians who are not so gloomy. They say "It is all because the intuitionists misunderstand the disjunction 'or'." These classical mathematicians say that "$A$ or $B$" means "$A$ is true or $B$ is true, but you don't have to know which one". That explanation doesn't sound unreasonable, but it has two disadvantages: first, "or" appears in the statement, and that is circular, and second, it is an explanation that gives rather little information. Anyone who uses a disjunction in daily life often wants to know more.

An instruction such as "to defuse the bomb you must cut the red wire or the green wire" is of little use to a novice bomb disposal technician, especially when the wrong choice will result in a regrettable explosion. For the said statement to be effectively correct, you must be able to identify one of the two disjuncts as the correct one.

Another case: when two authors C and M submit long, completely identical texts to a publisher, the latter will be inclined to say "C has copied, or M has copied". When asked for a justification, the publisher will say "It is impossible that neither copied from the other". But if at a later date a judge asks "who actually wrote this text?', there will be silence.[2] In logical terms, the publisher had claimed $A \vee B$, but gave as reason $\neg(\neg A \wedge \neg B)$.

So perhaps it is better to say that the traditional mathematician understands the disjunction $A \vee B$ as $\neg(\neg A \wedge \neg B)$, i.e., as "it is impossible that $A$ and $B$ are both false". This is a good definition, but what does it yield?

---

[2] The reader should perhaps be prepared not to go into exotic cases like Borges' Pierre Menard.

Consider another judge, who must hear the following case: in a hermetically closed room are found Smith, Jones and the body of Taylor. Taylor has been killed by a shot through the heart. A revolver with the fingerprints of Smith and Jones was found. Forensics showed that the deadly bullet came from this revolver and that suicide is out of the question. Neither Smith nor Jones confesses. If the judge uses the classical "or", he will say: "Smith or Jones is the murderer, because if neither of them is the perpetrator, Taylor would still be alive". But what is the informative value of this "or"? A conviction for murder cannot follow, because no shooter has been identified.

The moral is that whoever claims "$A$ or $B$" must be prepared to indicate the correctness of at least one of $A$ and $B$. In particular, whoever claims "$A$ or not-$A$" must be able either to show $A$ or to refute $A$. The reader will realize that our discussion in Chap. 1 about the presence or absence of a sequence of 20 consecutive nines in the decimal expansion of $\pi$ illustrates the non-constructive nature of PEM. Above, we saw that this is also the case for the Goldbach Conjecture. In fact, in the absence of a general method for solving mathematical problems, any as yet open problem is a weak counterexample to PEM!

Because Brouwer rejected the bivalence of logic, the truth table method was not going to be useful for intuitionistic logic. If you don't take the parts of a formula to be true, or to be false, how can you calculate a truth value of the whole?

It is an achievement of the first order that Brouwer managed to point the way to a coherent, systematic logic that is not truth-functional. He replaced the idea of *truth as correspondence with possibly unknown facts* by *truth as provability*. The latter he meant not in terms of the axioms and rules of a given formal system, but of mental constructions and of general ways of turning such constructions into other ones, of which we may come to see new ones in the future. Hence, the explanation of logic in intuitionism has become known as the *Proof Interpretation*. One also says that this is the intuitionist's *intended* interpretation of logic.[3] The ideas were found and used by Brouwer, but systematic presentation and work was first done by his former student Arend Heyting. The Russian mathematician Andrey Kolmogorov has given a closely related interpretation of constructive logic, which is why one speaks of the *Brouwer-Heyting-Kolmogorov Interpretation* or *BHK Interpretation*.

Brouwer's conception of logic originates in his conviction that the facts in mathematics exist because they have been brought about by our own constructive activity, so that a statement is proved by providing a mental construction. For example, you prove $2 + 3 = 5$ (in your mind) by constructing 2 and 3 and 5, performing the addition construction on 2 and 3, and then the construction showing equality with 5. The proof consists in the series of constructions as a whole. This is a proof in the basic sense. These are of course constructions in the intuitionistic sense: abstract constructions effected in our mind. What we write on paper and also call "proofs" are descriptions of the abstract proofs, and may serve to aid our memories or as instructions for others who want to carry out the same proof in their own minds.

---

[3] There are also unintended ones! One of them, the Kripke models, is discussed in Sect. A.2 of the Appendix.

The logical connectives serve to create compound statements, and are to be explained as operations on proofs in the basic sense. For the logical concepts the same is true as for any other: before a precise scientific definition can be given, one has an idea of what one is aiming for. Whether one wants to define "speed", "gravity" or "truth" in exact, unambiguous terms, one already has those concepts in mind in a vaguer way. In logic, the key concept is that of implication; "$A$ implies $B$" is among the most fundamental relations that our everyday and our scientific thinking can establish between an $A$ and a $B$. Brouwer saw that on his view of mathematics, implication must mean that we can carry out constructions that will get us from $A$ to $B$: if $A$ can be made evident, so can $B$. More precisely: "$A \rightarrow B$ is true" means that we have a construction method of which we see that, when applied to any proof of $A$, it will yield a proof of $B$. This can also be phrased in terms of an interaction between persons: "I have a construction that allows me to generate, whenever someone will come along and give me a proof of $A$, a proof of $B$ from it." That someone could be your future self!

Note that to be able to assert $A \rightarrow B$ thus understood, it is not required that we *have*, or know that we will have, a proof of $A$, and in cases we may even already know that there can be none because $A$ is false. Also, in practice we often have to reason from some hypothesis $A$ of which we have no idea whether in the end we will prove it or rather refute it. Heyting came up with a wonderful example [46]. At the time it was not known whether the sequence 0123456789 occurs in the decimal expansion of $\pi$.[4] But, Heyting pointed out, it can certainly be proved that "The sequence 0123456789 occurs in the decimal expansion of $\pi$" implies "The sequence 012345678 occurs in the decimal expansion of $\pi$". For if, by hypothesis, I have a construction for the longer sequence, I also have one for the shorter one.

Since there are no truth values, also $\bot$ has to be treated differently. It has to be defined as some concrete atomic statement for which it is immediately evident to us that an intended construction for it as some point "no longer goes" [17, p. 127]; For example, $1 = 2$. In fact, $\bot$ is just introduced for convenience, because in logic it is useful to be able to refer to a statement that is false without caring about its particular content.

Negation deserves some special attention. "$\neg A$ is true" must mean that the hypothesis that $A$ is true, i.e., that there is a proof of $A$, will lead to a falsehood, that is, to a proof of something that we know cannot be proved. So we define $\neg A := A \rightarrow \bot$. For if it is true that $A \rightarrow \bot$, i.e., if we have a construction method that converts a (hypothetical) proof of $A$ into a proof of $\bot$, then, since $\bot$ cannot be proved, $A$ cannot be proved either, hence cannot be true.

What about the converse—if $A$ has no proof, is there a construction method that converts a (hypothetical) proof of $A$ into a proof of $\bot$? The answer is "yes". Just take the identity operator $I$, which maps any given construction to itself. It satisfies the requirement, as we can safely promise that as soon as a proof $p$ of $A$ is presented, $I(p)$ will prove $\bot$. For that proof of $A$ will never come! More generally, if $A$ has no proof, then $A \rightarrow B$ is intuitionistically correct for any $B$. The argument is the

---

[4] As we mentioned in Chap. 1, with reference to [12], it does.

same. In particular, $\bot \to A$ holds for all $A$. The *Ex Falso* principle is therefore intuitionistically correct.[5]

Existence of a mathematical object can only mean, intuitionistically, that we have constructed it, or at least possess a method to construct it. This determines the meaning of $\exists x\, A(x)$, or, to make explicit that we quantify over the elements in a given domain $D$, $\exists x \in D\; A(x)$.

Universality takes the future into account. Just think of "Every human is an animal". This applies to humans living now, but also means that, whenever in the future an arbitrary human living then will be indicated, it is an animal. And similarly for past humans. What is expressed is part of the nature of humans.

A more mathematical example: when the mathematician is busy constructing natural numbers, and pauses to say "Any number is either even or odd", not only the numbers made so far are meant. For part of what is being said is "Whenever in the future I will have constructed some number, it will be either even or odd". Observe how universality therefore really expresses a kind of implication. So the meaning of $\forall x \in D\; A(x)$ will be: we have a construction method that, whenever we have a $d \in D$, can be applied to yield a proof of $A(d)$.

The future indicated by this "whenever" will be considered to be potentially infinite: after any mathematical construction, the subject can go on to make another one.

And what about the past? While at any moment the subject's past is finite, we make the idealization that it never forgets, so that all constructions it has made so far are present to its mind. In that sense, its present subsumes its past. So if at some point the subject is asked to give a proof of $A$, and in the past it has given such a proof, it is always available to it, however much time will have gone by.

Such idealizations lead to the idea of the *Creating Subject*, which is ideal in the same sense in which a Turing Machine is an ideal computer (can do whatever can be done in principle, has perfect memory, has endless time, never fails, etc.). We will return to the Creating Subject in Sect. 6.5.

We can neatly write out the Proof Interpretation. The $p$'s and $q$'s are always proofs, i.e., constructions, except in the case of the disjunction, where $p_0$ is a 0 or a 1, and the existential quantifier, where $p_0$ is a construction of whatever appropriate kind such that it has the property $A$. We see that also where the $p$ are not proofs, they are still constructions.

**Definition 2.5**  $p$ *is proof of* $P$ (notation "$p : P$") is, as far as the usual logical operators are concerned, explained as follows.

---

[5] This describes the common conception in current constructive practice. On the other hand, it fares badly with counterfactual situations. Imagine that it had been shown that there is no sequence 0123456789 in $\pi$. Then the described account supplies *a* reason why Heyting's example is just as much correct, but is wholly incapable of giving the reason most of us would give when asked: if by hypothesis I have a construction that produces a certain sequence of 10 decimals, then I certainly also have a construction that produces the first 9 of them. From the same perspective, there is an objection to *Ex Falso*, which presents us with a counterfactual by definition. For further discussion see, e.g., [86].

$$p : A \wedge B \qquad := p = (p_0, p_1) \text{ and } p_0 : A \text{ and } p_1 : B$$
$$p : A \vee B \qquad := p = (p_0, p_1) \text{ and } (p_0 = 0 \wedge p_1 : A) \text{ or } (p_0 = 1 \wedge p_1 : B)$$
$$p : A \rightarrow B \qquad := p \text{ is a construction that maps each } q : A \text{ to a } p(q) : B$$
$$p : \neg A \qquad := p \text{ is a construction that maps each } q : A \text{ to a } p(q) : \perp$$
$$p : \forall x {\in} D\ A(x) := p \text{ is a construction that maps each } a \in D \text{ to a } p(a) : A(a)$$
$$p : \exists x {\in} D\ A(x) := p = (p_0, p_1) \text{ and } p_0 \in D \text{ and } p_1 : A(p_0)$$
$$p : \perp \qquad := p : 1 = 2$$

Note:

(1) No definition is given for the proof of an atomic proposition. That would not be possible, because it depends on the particular objects and relations under discussion. The definition only deals with compounds, which lies in the very nature of logic.

(2) Constructively, if in fact or by hypothesis it is the case that $a \in D$, this means that we have, in fact or by hypothesis, a proof of this. Therefore, the construction $p(a)$ in the clause for $\forall$ can make use not only the object $a$, but also of that proof.

(3) Intuitionistically, there is no principled reason why these should be all logical operators! If logical operators are just highly general mathematical constructions, and mathematics is open-ended, then it cannot be excluded that further ones will be found.

Let us look at PEM again, now in the terms we have just acquired. $A \vee \neg A$ is true for an intuitionist if one can find a $p$ with $p : A \vee \neg A$, i.e., one finds a $p_0 \in \{0, 1\}$ and a construction $p_1$ such that $p = (p_0, p_1)$; furthermore, if $p_0 = 0$ then $p_1 : A$ and if $p_0 = 1$ then $p_1 : \neg A$. So one knows which of the two cases one is in, and depending on the value of $p_0$ one has a proof for $A$ or a proof for $\neg A$. In the latter case one knows that there is no proof for $A$, since every proof of $A$ is automatically converted into a proof for $\perp$, and $\perp$ has no proof.

Under this reading, PEM is exactly what has become known as *Hilbert's Dogma*, a conviction formulated by Hilbert in a famous lecture in Paris in 1900: for every mathematical statement $A$, one can either find a proof or show that it cannot have a proof. (So the principle is what we now call an *omniscience principle*.) He also presented the audience with a slightly different version, when he added that "in mathematics, there is no *ignorabimus*". He was referring to a then well known claim that the biologist Emil Du Bois Reymond (brother of the mathematician Paul) had made in 1872: when it comes to the questions what matter and force are, and how these can come to think, we must say *ignorabimus*—"we will never know". Hilbert's claim that in mathematics this is not the case amounts to claiming that in mathematics there are no statements that we can neither prove nor disprove. In other words: there are no absolutely unsolvable problems.

Brouwer already rejected Hilbert's Dogma in the first form in 1907,[6] but without identifying it with PEM. The reason lies in Brouwer's peculiar construal of PEM

---

[6] In the list of "Propositions" added to his dissertation, number XXI reads: "The following conviction of Hilbert [...] is unfounded: 'that every definite mathematical problem necessarily admits of rigorous settlement, be it that one succeeds in answering the question posed, be it that one demon-

at the time, conveyed by his remark that "'A function is either differentiable or not differentiable' says *nothing*; it expresses the same as the following: 'If a function is not differentiable, then it is not differentiable'" [28, p. 75].[7] In a paper of 1908, titled "The unreliability of the logical principles", Brouwer set things right, thereby introducing constructive logic.

However, the differently worded version of Hilbert's Dogma, according to which there are no absolutely unsolvable problems, is constructively correct. For suppose that we have an $A$ such that we can show that the assumption that it has been proved leads to a contradiction; according to the Brouwerian interpretation of the negation, as discussed above, this constitutes a proof of $\neg A$. So there cannot be such an absolutely undecidable $A$.

So PEM is not valid. However, Brouwer [18] showed that its double negation is correct. The argument he gives there is that if $\neg(A \vee \neg A)$ were correct, then so would $\neg A \wedge \neg\neg A$ be, but that is a contradiction, hence $\neg\neg(A \vee \neg A)$ is correct. (A formal derivation will be given in Appendix A.1.)

Similar reasoning is used in the following weak counterexample to $\neg\forall x \neg A(x) \rightarrow \exists x A(x)$ (Theorem 2.4(2) above). Let $R$ be the Riemann Hypothesis, and define

$$A(x) := (x = 1 \wedge R) \vee (x = 2 \wedge \neg R)$$

Now assume $\forall x \neg A(x)$, then in particular $\neg A(1) \wedge \neg A(2)$, hence $\neg R \wedge \neg\neg R$, contradiction. So $\neg\forall x \neg A(x)$. Yet until we have decided $R$, there can be no constructive proof of $\exists x A(x)$.

A useful little theorem proved by Brouwer is: $\neg A \leftrightarrow \neg\neg\neg A$ [23]. Again we will give the argument informally, just as Brouwer did. To start with, $A \rightarrow \neg\neg A$. For, assume $p : A$ and $q : \neg A$, then by definition $q(p) : \bot$ holds. That is, the construction $r$ with $r(q) = q(p)$ satisfies $r : \neg A \rightarrow \bot$, or $r : \neg\neg A$. Now for $s$ with $s(p) = r$ we have that $s : A \rightarrow \neg\neg A$. If we now substitute $\neg A$ for $A$, we get that $\neg A \rightarrow \neg\neg\neg A$. For the other direction, suppose we have a proof of $\neg\neg\neg A$, that is, of $\neg\neg A \rightarrow \bot$. Combining the latter with $A \rightarrow \neg\neg A$ from above, we get $A \rightarrow \bot$, i.e., $\neg A$.

## 2.3  Intuitionistic Logic is Not Three-Valued

One early reaction to intuitionistic logic was: If one does not accept PEM, one must accept that there are propositions that have a truth value that is neither true, nor false,

---

strates the impossibility of its solution and thereby the necessity of the failure of all attempts'." [28, p. 101].

[7] The original is [17, p. 131]. This construal can be found in a book by Bellaar-Spruyt (1842–1901), the teacher of logic in Amsterdam, whose course Brouwer probably took; see [96, pp. 106–107].

but "third" [2, p. 59].[8] Notation: $A'$ for "$A$ is third".[9] A theorem of Valery Glivenko [40] showed that this is not the case.

His intuitionistic theorem, arrived at in step (7) below, has a general applicability: it means that if a proposition can be proved false in classical propositional logic, it can be proved false in intuitionistic propositional logic. Glivenko's proof, in compacted form:

$$
\begin{aligned}
&(1)\ (A \vee \neg A) \to \neg B &&\text{assumption}\\
&(2)\ \neg\neg(A \vee \neg A) \to \neg\neg\neg B &&\text{contraposing (1) twice}\\
&(3)\ \neg\neg\neg B \to \neg B &&\text{theorem (Brouwer)}\\
&(4)\ \neg\neg(A \vee \neg A) \to \neg B &&\text{(2), (3), and transitivity of } \to\\
&(5)\ \neg\neg(A \vee \neg A) &&\text{theorem (Brouwer)}\\
&(6)\ \neg B &&\text{(4), (5), and modus ponens}\\
&(7)\ ((A \vee \neg A) \to \neg B) \to \neg B &&\text{(1), (6), and } \to\text{-introduction}
\end{aligned}
$$

Application to "third":

$$
\begin{aligned}
&(8)\ A \to \neg A' &&\text{true excludes third}\\
&(9)\ \neg A \to \neg A' &&\text{false excludes third}\\
&(10)\ (A \vee \neg A) \to \neg A' &&\text{(8), (9), and propositional logic}\\
&(11)\ \neg A' &&\text{(7) with } A' \text{ for } B, \text{ (10), and modus ponens}
\end{aligned}
$$

Since $A$ here is arbitrary, this shows that in intuitionistic logic no proposition is "third", thereby refuting Barzin and Errera's claim.

In a comment on Glivenko's proof, Barzin and Errera [3, p. 13] state that the meaning of $\neg\neg(A \vee \neg A)$, which Glivenko proves as a lemma and is used in step (5) above, can only be: "There exists no proposition that is third". They then argue: if $A \vee \neg A$ is not false, and Brouwer holds that it is not true either, then it must be third; contradiction.

This calls for two remarks. First, as we have seen above, to intuitionists truth values are not the primitive notion used in explaining logic. For them, that is rather *proof*.[10] They also allow *forming the hypothesis* that one has obtained a proof. In these terms they obtain an explanation of the meaning of $\neg\neg(A \vee \neg A)$ without reference to the existence or non-existence of any truth value, and on which that proposition is furthermore correct. Second, when Brouwer says that PEM is not true, this is a weak negation: he means the current absence of a proof of PEM, not the presence of a refutation.

---

[8] For the historical details of this episode, see [45].

[9] In fact, Barzin and Errera themselves denied that a third truth value exists, and therefore took their argument that intuitionistic logic is three-valued to be an argument against intuitionistic logic.

[10] Or, in Kolmogorov's similar explanation, *solution to a problem*.

## 2.4  Indirect Proof

There is an age-old technique of proving theorems in a roundabout way: you first assume the opposite of what you want to prove and then reason until you find a contradiction. Then you conclude that the statement to be proved is correct. Euclid, for example, used this method all the time. We give two examples of it.

(1) From a point $P$ outside a line $l$ only one perpendicular can be drawn. Suppose that two different perpendiculars can be drawn from $P$, say to points $Q$ and $R$ on $l$, then a triangle with vertices $P, Q, R$ is created. Since both angles at $Q$ and $R$ are right, the sum of the angles is now greater than $180°$. But that contradicts the theorem of Euclidean geometry that in any triangle the sum of the angles is $180°$. Ergo: the two perpendiculars coincide.[11]

(2) If a rational number has no inverse, then it is 0. Let $p/q$ be a rational number with no inverse. Suppose $p/q \neq 0$. Then $p \neq 0$. So the rational number $q/p$ exists, and $(p/q)(q/p) = 1$. So $p/q$ has an inverse. Contradiction. Ergo: $p/q = 0$.

Now let us see what is possible according to this method, and why. We want to show a certain statement $A$; now we take $\neg A$ and derive a contradiction. According to Brouwer's explanation of the negation, thereby $\neg\neg A$ has been proved. To get $A$ we would need an implication $\neg\neg A \to A$. Classically that is no problem, the idea being that "There are two possibilities, $A$ is true or $\neg A$ is true. So if $\neg A$ is false, $A$ is true."

But such appeal to there being exactly two possibilities is in fact an appeal to PEM, and constructively there is no evidence for that as a general principle; the double negation principle is essentially equivalent to proof by contradiction. So there seems to be little to argue in favor of proof by contradiction. Intuitionists therefore accept indirect reasoning as a proof that $\neg A$ is false, but refuse to go on and take that as a proof that $A$ is true. (We will see another example in this chapter, when we discuss the question whether there are infinitely many primes.)

Because proof by contradiction is so useful, few people are willing to give it up. However, it must be admitted that such a proof provides too little information. Here is an example that shows that while proof by contradiction works wonders, from a constructive point of view it promises more than it delivers.

**Theorem 2.6** *There are two irrational numbers $a$ and $b$ such that $a^b$ is rational.*

**Proof** Suppose this is false. Then all powers $a^b$ of irrational numbers are irrational. Let us try one of the oldest known irrational numbers: $\sqrt{2}$. Set $a = \sqrt{2}, b = \sqrt{2}$. By our assumption, $a^b$ is an irrational number. But if we raise that number to the power $\sqrt{2}$, we get $(\sqrt{2}^{\sqrt{2}})^{\sqrt{2}} = (\sqrt{2})^{\sqrt{2}\sqrt{2}} = (\sqrt{2})^2 = 2$. This contradicts the assumption. Conclusion: there are indeed two such irrational numbers.  □

---

[11] With geometric proofs one has to be very careful—what can and what cannot be used? Axiomatic theories can be intricate artifacts. What we have used here can be found in Euclid.

What does this proof actually give us? Are we now able to indicate an example of such a pair of $a$ and $b$? The candidates are the pairs $\langle \sqrt{2}, \sqrt{2} \rangle$ and $\langle \sqrt{2}^{\sqrt{2}}, \sqrt{2} \rangle$, but which of the two is the right one, the above argument does not say. What we in fact showed is that we cannot obtain an irrational number in both cases. The sober commentary is "Is that all?" On the constructive view, whoever claims that some mathematical object with a certain property exists is obliged to produce that object if requested. Well, on the basis of the above proof by contradiction that is not possible. *Moral*: Proof by contradiction does not always yield a constructive result.

**Excise 2.7** We could have used the Principle of the Excluded Middle directly instead of proving by contradiction. Provide that proof yourself.

The example also serves to show that the classical disjunction is not effective.

**Excise 2.8** For certain other pairs of irrational numbers we do get constructive results.

(1)  Show that $a = \sqrt{2}$ and $b = \log_2 9$ form a good pair.
(2)  Also, $a = \pi, b = \log_\pi 2$; recall that $\pi$ is transcendental.

Proof by contradiction is nevertheless useful, namely where we aim to derive a negative fact. Indeed, we just used one:

**Theorem 2.9**  $\sqrt{2}$ *is irrational.*

Since "irrational" means "not rational",[12] you are asked to show that something *is not true*: there are no natural numbers $p$ and $q$ such that $\sqrt{2} = p/q$. According to our explanation of the meaning of the negation, we must therefore derive a contradiction from $\exists p \exists q (\sqrt{2} = p/q)$. The old proof is as clever as it is simple.

***Proof*** To begin with, we may simplify the fraction $p/q$ so that the numerator and denominator have no common factor. Now if $\sqrt{2} = p/q$, then $2 = p^2/q^2$. So $2q^2 = p^2$. The right-hand side must therefore be divisible by 2. Therefore, $p^2$ is even, and so $p$ must be even (see the exercise below). But then the right-hand side is divisible by 4, and so is the left-hand side. Then $q^2$ is also divisible by 2, so $q$ is also even. Now $p$ and $q$ are both divisible by two. Contradiction. Conclusion: the requested $p$ and $q$ do not exist, i.e., $\sqrt{2}$ is irrational. $\qquad\qquad\square$

In this book, we will not get too formal. We simply reason effectively, and logic comes to the rescue when clarification is needed, or when there is something nice to demonstrate. We will see an example in Sect. 3.1: the *Least Number Principle*.

---

[12] Another (more demanding) definition of irrationality would be that an irrational number is one that lies *apart* from every rational number; see Definition 4.27. This is in fact called *strong irrationality*. We will come back to it at the end of Sect. 6.7.

## 2.5  A Concluding Remark

In this chapter we have seen that Brouwer's ideas about existence and truth in pure mathematics led him to conclude that the logic now known as classical logic—and when Brouwer began his investigations there was no other—was not correct, and makes many false suggestions. As Brouwer came to argue in a paper the year after his dissertation, PEM is "unreliable", whereas constructive logic does not suffer from that defect [18]. Although this improved logic for him certainly had instrumental value, he emphatically did not accord it any significance for the foundations of mathematics. But that is because he also held that logic, of whatever kind, is essentially different from mathematics. A few weeks before his thesis defense, he wrote in a letter to his supervisor Diederik Korteweg:

> In the beginning of the chapter,[13] I show that mathematical reasoning is *not* logical reasoning, and that it is just out of poverty of language that mathematical reasoning uses the connectives of logical reasoning; [· · ·] Far from it being "strange people" that do not argue logically, I do believe that it's just a phenomenon of inertia that the corresponding words still exist in modern languages. Pure usage of those words only rarely occurs, and impurely they are used in daily life, where they have led to all kinds of misunderstandings and dogmatism, and in mathematics to the misconceptions of set theory. Those misconceptions arose *not* through insufficient mathematical insight, but because mathematics, lacking a pure language, has *to make do with the language of logical reasonings*, whereas *its thoughts reason not logically, but mathematically,* which is something completely different.

And further on:

> Theoretical logic doesn't teach anything in the present world, and people know this, at least the sensible ones; it serves only lawyers and demagogues anymore, not to instruct other people but to deceive them. And that that is possible, is because the vulgus unconsciously reasons: "that language with logical figures exists, so it will probably also be useful", and so they meekly let themselves be deceived; just as I have heard several people defend their gin-drinking with the words: "What else is gin for?" Whoever has illusions to improve the world can with equal justification agitate against the language of logical arguments as against alcohol; no more is it a "strange people" that does not drink alcohol, than it is a "strange people" that does not reason logically; although I believe that perhaps no abuse is more entrenched than what has thus grown together with the most common parts of the language. [100, pp. 38, 39, translation modified]

Not every supervisor would know what to do with the somewhat rebellious language of a Ph.D. student—but Korteweg let Brouwer open the way to intuitionism. It is in any case clear that, whether or not one shares Brouwer's view on the relation between mathematics and logic, an emphasis on logic hinders intuitionistic mathematics in coming into its own.

---

[13] Chapter 3 of his 1907 dissertation.

# Chapter 3
# Natural Numbers, Integers, and Rationals

**Abstract** We first construct the natural numbers, in a way that, according to Brouwer, is based on the fundamental intuition of the passage of time. This intuition justifies definition by recursion and the Principle of Complete Induction. We also illustrate the difference between effective and non-effective proofs; for example, the Least Number Principle is not constructively valid. Equality on the natural numbers is shown to be decidable; and that property extends to the integers and rational numbers, which we construct as equivalence classes. We discuss the orderings, in particular of the rational numbers.

## 3.1 Natural Numbers

Now it is time to begin constructing mathematics. Obviously, this chapter and this book can only deal with a fragment. The natural numbers are a good place to start, as we are highly familiar with them from everyday life, and practically all the theoretical development of mathematics presupposes them. In our language, we will be using intuitionistic logic as introduced in the previous chapter, without going into technical or philosophical details. But we will have to be careful not to use non-constructive principles.

Following Brouwer, we take the position that the natural numbers are the results of acts of creation in the mind, based on a fundamental intuition of the passage of time from one moment to another.[1] That intuition provides content to the idea that each number $n$ has a successor, $S(n)$. We call $S$ the *successor function*. We are not concerned here about whether the natural numbers in their natural order start at 1 or at 0, but it seems to be most efficient to start with 0. Since intuitive time's arrow moves in one direction, no number created on the basis of the intuition of the passage of time can appear twice: $0 \neq S(x)$ and $S(x) = S(y) \to x = y$.

---

[1] We will return to this at the beginning of Chap. 4, and look at some of Brouwer's own formulations of how the mental construction of mathematics begins in Chap. 6.



D. van Dalen et al., *Intuitionistic Analysis*, Springer Undergraduate Mathematics Series,
https://doi.org/10.1007/978-3-032-16491-9_3

Any number $n$ that is not 0 can be given to us as a construction $S(S(\ldots S(0))\ldots)$. This is sometimes called its *canonical form*. Usually natural numbers are given to us in a non-canonical form, such as $2 + 2$, or 4, instead of $SSSS0$.

We accept that we can perform an unbounded iteration of the operation of creating the successor.

This mental construction of the collection of natural numbers, one element after another, has two important consequences—proof by induction and definition by induction, more commonly called *definition by recursion*. We discuss the latter first. Simply put, an operation $f(x, y, z)$ is defined recursively by following the construction of the natural numbers for one of the variables. This is how we will obtain operations such as addition, multiplication, exponentiation, and cut-off subtraction.

**Definition 3.1** (*Sum*)

$$n + 0 = n$$
$$n + S(m) = S(n + m)$$

Thus one only needs to know the successor operation, and how to handle a recursive definition, to be able to determine the sum of two numbers.

We will use the traditional notation for natural numbers from now on:

$$1 = S(0), 2 = S(1), 3 = S(2), 4 = S(3), 5 = S(4), \ldots$$

We will also write $n + 1$ for the successor $S(n)$.

Now let us prove as *proof of the pudding* that $2 + 2 = 4$. According to our convention, $2 = S(S(0))$. We now calculate $2 + 2 = S(S(0)) + S(S(0)) = S[S(S(0)) + S(0)] = S(S([S(S(0)) + 0])) = S(S(S(S(0)))) = S(S(S(1))) = S(S(2)) = S(3) = 4$.[2] There is nothing more to add. Based on our intuition of "number" and "addition", we have conclusively established that $2 + 2 = 4$.

Multiplication is determined according to the same principle: reduce the calculation to a smaller input.

**Definition 3.2** (*Product*)

$$n \cdot 0 = 0$$
$$n \cdot S(m) = n \cdot m + n$$

We follow the convention that, if no confusion arises, the multiplication sign is omitted, i.e., we write $xy$ for $x \cdot y$.

**Exercise 3.3** Now also prove $2 \cdot 2 = 4$.

---

[2] The large numbers of $S$'s in a row do not deserve a prize for beauty. The notation is agreed upon for executing functions sequentially. Think of $\log(\log(10)) = \log(1) = 0$ and $f(g(x))$. When repeatedly applying the same function, the " power notation" is applied: $f(f(f(x))) = f^3(x)$. Since we are only dealing with short iterations here, we forego the use of such abbreviations.

Another convention: As is customary, and as we did in logic in Chap. 1, we will be a bit economical with parentheses. For starters, we omit the parentheses from the successor when no confusion is possible. For example, we write $3 = SSS(0)$. Furthermore, we use the common precedence convention: exponentiation > multiplication > division > taking square roots > addition > subtraction.

Recursive definition yields a huge stock of functions. Here are some more:

**Definition 3.4** (*Exponentiation*)

$$n^0 = 1$$
$$n^{S(m)} = n^m \cdot n$$

**Definition 3.5** (*Factorial*)

$$0! = 1$$
$$S(n)! = n! \cdot S(n)$$

**Definition 3.6** (*Predecessor*)

$$\mathrm{pd}(0) = 0$$
$$\mathrm{pd}(S(n)) = n$$

**Definition 3.7** (*Cut-off subtraction, "monus"*)

$$x \mathbin{\dot-} 0 = x$$
$$x \mathbin{\dot-} S(n) = \mathrm{pd}(x \mathbin{\dot-} n)$$

Since we have taken 0 as the starting point of the sequence of natural numbers, it actually has no predecessor, but that is okay. We simply say that it is its own predecessor. In mathematics, you may deviate from ordinary language if you have a good reason and warn the reader.

**Definition 3.8** (*Minimum*) $\min(x, y) = x \mathbin{\dot-} (x \mathbin{\dot-} y)$

**Definition 3.9** (*Maximum*) $\max(x, y) = x + (y \mathbin{\dot-} x)$

**Definition 3.10** (*Summation*)

$$\sum_{z=0}^{z=0} f(z) = f(0)$$
$$\sum_{z=0}^{z=x+1} f(z) = \sum_{z=0}^{z=x} f(z) + f(x+1)$$

**Definition 3.11**  (*Product*)

$$\prod_{z=0}^{z=0} f(z) = f(0)$$

$$\prod_{z=0}^{z=x+1} f(z) = \prod_{z=0}^{z=x} f(z) \cdot f(x+1)$$

We also write

$$\sum_{z<x} f(z) \text{ for } \sum_{z=0}^{z=x-1} f(z) \text{ and } \prod_{z<x} f(z) \text{ for } \prod_{z=0}^{z=x-1} f(z),$$

and for the sake of completeness we suggest setting

$$\sum_{z<0} f(z) = 0 \text{ and } \prod_{z<0} f(z) = 1$$

**Exercise 3.12**

(1)  Using these definitions, calculate $2^3$, $4!$, $\mathrm{pd}(5)$.
(2)  Define the signum function sg, which satisfies

$$\mathrm{sg}(x) = 0 \text{ if } x = 0$$
$$\mathrm{sg}(x) = 1 \text{ if } x > 0$$

(3)  Define "the distance between x and y", i.e., $|x - y|$, using cut-off subtraction.

We note that functions $f$ defined by recursion are always *computable*, i.e., that for $f$ there is an algorithm that calculates the output $f(n)$ for each input $n$.

In addition to *defining by reducing to predecessors*, which is what recursive definitions amount to, there is also a method of *proving by reducing to predecessors*. This is called the *Principle of Complete Induction*. In words: If a property holds for 0, and from the validity for a number follows that for its successor, then it holds for all natural numbers:

$$(A(0) \wedge \forall x (A(x) \rightarrow A(x+1))) \rightarrow \forall x A(x)$$

About the notation for variables: The intuitive role of variables is that they represent arbitrary elements of the domain that the statements are taken to be about. A traditional formulation for this is "the variable $x$ runs over …". Which letters of the alphabet are chosen for variables is not so important. There are habits and conventions for such matters, but we will not worry about it so much here. In any case, it is an old custom to indicate integers with the letters $k, l, m, n$, with or without an index. Pay close attention to what the letters stand for.

Possibly the simplest example of the use of  induction is the justification of recursion : suppose $f$ is defined by

$$f(n) = g(n)$$
$$f(S(n)) = h(f(n)),$$

where $g, h$ are functions already given. To show that, for all $x$, $f(x)$ is well-defined, is to show that whenever we have constructed an $x$ in the domain of $f$, here $\mathbb{N}$, application of the construction method specified in the definition of $f$ succeeds and yields an object in the range of $f$, here also $\mathbb{N}$. In the present case, note first that

$$g(0) \text{ is defined } \wedge f(0) = g(0) \to f(0) \text{ is defined.}$$

And,

$$(\text{for all } x, h(x) \text{ is defined}) \wedge f(n) \text{ is defined} \to h(f(n)) \text{ is defined}$$
$$\to f(S(n)) \text{ is defined.}$$

By induction, for all $n$, $f(n)$ is defined.

The  Principle of Complete Induction[3] is the most natural thing in the world. The idea is that we build the natural numbers and make proofs at the same time.

| | | |
|---|---|---|
| Step 0. | Take the number 0; | prove $A(0)$. |
| Step 1. | Go from 0 to 1; | use $A(0)$ and $A(0) \to A(1)$ to prove $A(1)$. |
| $\vdots$ | $\vdots$ | $\vdots$ |
| Step $n + 1$. | Go from $n$ to $n + 1$; | use $A(n)$ and $A(n) \to A(n + 1)$ to prove $A(n + 1)$. |

This procedure provides a proof for any desired number.[4]

Note that to prove $\forall n\, A(n)$ you only need to prove $A(0)$ and $A(n) \to A(n + 1)$. In the latter implication, $A(x)$ is called the *induction hypothesis*.

---

[3] In colloquial speech, induction refers to the transition from a limited number of experiences to an entire class. If you've been on holiday in Crete, and you've only met people who lie, you come back and tell your friends that "all Cretans lie". And if you have only become acquainted with the house and garden birds in our country, which all fly, then it is obvious to say "all birds fly". This type of induction is not very reliable. In mathematics, we do not wish to reason this way. Whoever regularly adds two natural numbers will observe that their order does not matter for the result. Nevertheless, a mathematician does not want to say that therefore in adding two natural numbers, their order does not matter. One wants to see proof first. It can often be given with the  Principle of Complete Induction. The adjective "complete" indicates that we don't just go from a number of samples to the conclusion, but that we follow a procedure that indeed deals with *all* natural numbers.

[4] For philosophical and historical discussion of the justification of complete induction, as well as further references, see [91].

**Exercise 3.13** Prove by complete induction that every natural number is even or odd.

The Principle of Complete Induction is the most important tool for establishing properties of natural numbers. There is an alternative form of this principle which is often useful in applications:

**Definition 3.14** (*Least Number Principle*) If there is an element $n$ with the property $A(n)$, then there is a smallest element with that property. In symbolic form this reads

$$\exists x\, A(x) \rightarrow \exists x (A(x) \wedge \forall y < x \neg A(y))$$

In classical mathematics, the Least Number Principle is equivalent to the Principle of Complete Induction; the proof is not very difficult—try to prove the equivalence yourself. The principle is often used in negative form: if $A(x)$ somewhere goes wrong, there is a smallest $n$ for which it goes wrong (Principle of the Smallest Counterexample).

In intuitionistic mathematics the principle does not work. After all, the obvious thought is, start with the given number $n$ for which $A(n)$ holds and then test all predecessors. But testing all predecessors is only possible if the property $A(x)$ is decidable, that is, for each $m$ it can be determined whether $A(m)$ or $\neg A(m)$. And without PEM we cannot claim such a thing. It is not difficult to devise a weak counterexample. Consider the predicate

$$A(x) := x = 5$$
$$\vee\ (x = 1 \ \wedge \ \text{there is a sequence of 20 nines in } \pi)$$
$$\vee\ (x = 3 \ \wedge \ \text{there is no sequence of 20 nines in } \pi)\,.$$

In plain language this formula says that 5 has the property $A$ and if there is a sequence of 20 nines in $\pi$ then 1 has the property and otherwise 3 has the property. Obviously, $\exists x\, A(x)$. Also, if there were a smallest element $n$ with the property $A$, it couldn't be 5. (Why not? What would have to be the case if it were?) But we cannot now say that 1 or 3 has the property. So a smallest element cannot now be constructed. *Conclusion*: the Principle of the Smallest Element is not constructively valid.

Note that this means that constructively the natural numbers are not well-ordered in the sense of classical set theory: "An ordered set is well-ordered if every non-empty subset has a first element". There are alternative constructive and intuitionistic ideas about well-ordering; apart from a brief remark in Sect. 6.1, we will not go into that subject in this book. Here we just note that the natural ordering of the natural numbers is considered to be a well-ordering because it can be defined inductively.

**Exercise 3.15**

(1)  Prove $S(\mathrm{pd}(S(n))) = S(n)$ for all $n$. Does $S(\mathrm{pd}(n)) = n$ also hold?

If a formula contains more variables, you have to be a bit clever and choose the most useful variable for the induction.

(2)  Prove $(x + y) + z = x + (y + z)$ by induction on $z$. Why are $x$ and $y$ not so useful?

(3)  Prove $0 + x = x$ and $S(x) + y = y + S(x)$. Using this prove $x + y = y + x$.

The above seems to suggest that we are quite capable of defining all kinds of functions using recursion, but would that be all? Can we, for example, also introduce relations that have the same effective nature? The answer is yes; we can introduce numerous relations that are functions in disguise. Here are two examples:

**Definition 3.16**

$$x \leq y := x \mathbin{\dot{-}} y = 0$$
$$x < y := S(x) \mathbin{\dot{-}} y = 0$$

The functions and relationships we introduced above are only a small part of what is used in practice. The theory of arithmetic with recursively defined functions and relations is, together with a  weak form of induction, called *primitive recursive arithmetic*, PRA.[5] This theory plays an important role in logic. It is useful for us to consider that the functions of PRA can be computed in an easy, well-arranged manner.

An important property of arithmetic is that  equality is  decidable, i.e.,

**Theorem 3.17**  $\forall x \forall y(x = y \vee x \neq y)$

In words: PEM holds for equality of natural numbers.

There are two ways of convincing oneself of this. The first is to take it as evident upon understanding it: the construction of a number is ultimately a process of adding units. If I begin the construction of $x$ and of $y$ at the same time, and take the steps in their construction processes in parallel, then either both processes end simultaneously, or one of them finishes before the other.

The other way is to prove this from more elementary properties and relations between the numbers, in order to honor the tradition of choosing the simplest axioms necessary. To some the following proof may seem excessively formal and detailed. But one should realize that now we cannot use direct insights into processes. (Such insight is nevertheless present implicitly, because the acceptance of the Principle of Complete Induction depends on it.)

**Lemma 3.18**  $S(0) = 0$ *behaves as a falsum, i.e., from* $S(0) = 0$ *everything follows.*

*Proof*

(1)  $S(0) = 0 \to x = 0$. Induction on $x$:

  (a)  *Basis.* If $x = 0$ the implication evidently holds, since $0 = 0$.

---

[5] PRA is a quantifier-free theory, i.e., only free variables play a role (whence the  "weakness" of the induction available—only for quantifier-free formulas). Nevertheless, the theory is surprisingly flexible. See also [83, Sect. 2.1].

(b) *Induction step.* Assume $S(0) = 0 \to x = 0$. But then

$$S(0) = 0 \to S(x) = S(0)$$
$$\to S(x) = 0\,.$$

We conclude that $\forall x\bigl(S(0) = 0 \to x = 0\bigr)$.

(2) $S(0) = 0 \to x = y$, since, by (i), $x$ and $y$ both equal 0 if $S(0) = 0$.

Thus, we see that all atoms follow from $S(0) = 0$. Then it is a matter of routine (for those who can give inductive proofs on formulas) to show that $S(0) = 0 \to A$ for all formulas $A$. This means we can define $\bot$ as $1 = 0$. $\qquad\square$

**Proof** (of Theorem 3.17) The proof of $\forall x \forall y(x = y \vee \neg x = y)$ is a double induction. First, we rewrite this as
$$\forall y \forall x(x = y \vee \neg x = y)\,.$$

Now we will prove $\forall x(x = 0 \vee \neg x = 0)$ by induction on $x$, and then use that as the basis for a proof of $\forall y \forall x(x = y \vee \neg x = y)$ by induction on $y$.

(1) *Basis.* $\forall x(x = 0 \vee \neg x = 0)$.

    (a) *Basis.* Clearly $0 = 0$ and $0 = 0 \to 0 = 0 \vee \neg 0 = 0$.

    (b) *Induction step.* We have already remarked that $\neg S(x) = 0$, whence $S(x) = 0 \vee \neg S(x) = 0$, i.e.,

$$x = 0 \vee \neg x = 0 \to S(x) = 0 \vee \neg S(x) = 0\,,$$

    and induction yields $\forall x(x = 0 \vee \neg x = 0)$.

(2) *Induction step.* Let $y$ be given and assume $\forall x(x = y \vee \neg x = y)$. We wish to derive $\forall x(x = Sy \vee \neg x = Sy)$ from this. To this end, let $x$ be arbitrary. By (1), $x = 0 \vee \neg x = 0$ and we can argue by cases.

    (a) $x = 0$. Observe

$$x = Sy \to 0 = Sy$$
$$\to Sy = 0\,.$$

    Thus, $\neg 0 = Sy$, i.e., $0 = Sy \vee \neg 0 = Sy$.

    (b) $\neg x = 0$. We need two additional facts we will prove shortly:

        (i) $\forall z \forall w(z = w \leftrightarrow Sz = Sw)$

        (ii) $\forall z(\neg z = 0 \to \exists w(z = Sw))$.

    By (ii), we can write $x = Sw$. By the induction hypothesis, $w = y \vee \neg w = y$. Now, by (i), $Sw = Sy \leftrightarrow w = y$. Thus,

$$w = y \rightarrow Sw = Sy$$
$$\rightarrow x = Sy$$
$$\rightarrow x = Sy \vee \neg x = Sy\,,$$

and

$$\neg w = y \rightarrow \neg Sw = Sy$$
$$\rightarrow \neg x = Sy$$
$$\rightarrow x = Sy \vee \neg x = Sy\,.$$

In either case, $x = Sy \vee \neg x = Sy$. But $x$ was arbitrary, so $\forall x(x = Sy \vee \neg x = Sy)$.

By induction, it now follows that $\forall y \forall x(x = y \vee \neg x = y)$.

To complete the proof, we only need to give the two proofs we had postponed. Proof of (i). The implication $x = y \rightarrow Sx = Sy$ is logically valid. Conversely,

$$Sx = Sy \rightarrow \mathrm{pd}(Sx) = \mathrm{pd}(Sy)\,, \quad \text{again by logic,}$$
$$\rightarrow x = y\,, \quad \text{by definition of } \mathrm{pd}\,.$$

Proof of (ii). By induction.

(a) *Basis.*

$$\neg 0 = 0 \rightarrow \neg 0 = 0$$
$$\rightarrow (0 = 0 \rightarrow \bot)$$
$$\rightarrow (0 = 0 \rightarrow \exists y(0 = Sy))$$

But $0 = 0$, whence $\neg 0 = 0 \rightarrow \exists y(0 = Sy)$.

(b) *Induction step.* Note that $Sx = Sx$, whence $\exists y(Sx = Sy)$, whence $\neg Sx = 0 \rightarrow \exists y(Sx = Sy)$.

$\square$

The decidability of equality for natural numbers can be extended without further ado to integers and rational numbers, which, as we will see in the next section, can be constructed out of the natural numbers. The difficulties only start with the real numbers, as we saw in Chap. 1.

A motto of Brouwer's mathematics could be: "Show and tell!". Here is a simple example. Claim: There is a prime number between 10,000 and 10,100. One thing you could so is to take a prime number table and find the first prime number after 10,000 (you can also search the internet). Is a mathematician satisfied now? Not really, one needs to check if the number found, 10007, is really prime—the creator of the table might have made a mistake. Fortunately, we know how to check whether a number is prime, and we're done.

Now a more complicated, and also more interesting example.

**Theorem 3.19**  *For every prime number there is a greater one.*

***Proof***  Suppose it is false. Then there are only finitely many primes, $p_0, p_1, p_2, \ldots,$ $p_n$. Now consider the number $a = p_0 p_1 p_2 \ldots p_n + 1$. We see that $a$ is not divisible by a prime number. But there is a theorem that any number can be factored into prime numbers. Contradiction. So the assumption that there is a last prime number cannot be true. Ergo, after each prime number there must be yet another prime number.  $\square$

We have hereby given a proof by contradiction, as discussed in Chap. 2. Such a proof takes the opposite of what you want to prove and infers a contradiction. If the opposite of what you want to prove cannot be true, then what you want to prove must be true. But does this really answer the question? If something cannot possibly be false, what is it then: true, or in limbo? Does the above proof deliver an infinite sequence of prime numbers? That depends on what we mean by infinite. There are two answers: (1) The sequence cannot be finite, (2) We can extend any finite sequence.

The argument just given proves the first claim, but not the second. We have not shown that after $p_0, p_1 \ldots p_n$ another $p_{n+1}$ follows, i.e., we did not find the claimed $p_{n+1}$. We have only shown that it is impossible that there be no $p_{n+1}$. Fortunately, we can give a constructive proof (essentially Euclid's).

***Proof***  To show that after every prime number $p_n$ a greater prime can be found, we assume that $p_0, p_1, \ldots, p_n$ have already been found. Look again at the number $a$ from above. Because $a$, by the general theorem, is a product of primes, and since $a$ is not divisible by $p_0, \ldots, p_n$, it must be divisible by a prime number $p$ greater than $p_n$, but of course not greater than $a$. We are now going to factorize $a$, and then we will find a prime $p$ that is greater than the ones we had.  $\square$

Of course, that prime $p$ doesn't have to be the first prime number after $p_n$.

**Exercise 3.20**  How do we now find the first prime number after $p_n$?

The above clearly illustrates the difference between effective and non-effective; the difference between being able to calculate and saying "it is there, but I don't know where".

## 3.2   Integers and Rational Numbers

The fundamental number types of mathematics are the natural numbers, integers, rational numbers, real numbers, and complex numbers. Real numbers are the topic of the next chapter, whereas complex numbers will not be treated in this book.

Obtaining the integers from the natural numbers and then the rational numbers from the integers is constructively unproblematic.

Integers are constructed as pairs of natural numbers. Heuristic: when we think of $(m, n)$, we think of $m - n$, and that last number can turn out to be positive, zero or negative.

**Definition 3.21** Two pairs of natural numbers $(m_1, n_1)$, $(m_2, n_2)$ are *equivalent* if $m_1 + n_2 = m_2 + n_1$. Notation: $(m_1, n_1) \sim (m_2, n_2)$.

**Exercise 3.22** Verify that $\sim$ is an equivalence relation. What does this equivalence mean for the heuristic just given?

**Definition 3.23** The operations $+$, $\times$ and $-$ on such pairs are defined as follows:

$$(m_1, n_1) + (m_2, n_2) := (m_1 + m_2, n_1 + n_2)$$
$$(m_1, n_1) \times (m_2, n_2) := (m_1 m_2 + n_1 n_2, m_1 n_2 + m_2 n_1)$$
$$-(m, n) := (n, m)$$

The operation signs on the left are the same as those on the right, but the ones on the right are interpreted as those defined on the natural numbers.

**Exercise 3.24** Show that these operations preserve equivalence.

**Definition 3.25** $(m_1, n_1) < (m_2, n_2) := m_1 + n_2 < m_2 + n_1$

**Exercise 3.26** Show that the ordering is preserved under equivalence.

**Definition 3.27** The equivalence classes of pairs under $\sim$ are called *integers*.

In practice, we write $(m, n)/_\sim$ as $m - n$. Example:

$$(0, 4)/_\sim = (7, 11)/_\sim = \cdots = -4$$
$$4 + (-4) = (4, 0)/_\sim + (0, 4)/_\sim = (4, 4)/_\sim = (0, 0)/_\sim = 0 \,.$$

Note that, in the informal representation, $m - 0$ yields exactly the embedding of the natural numbers in the integers. A wider notion of number has been introduced, and then you want to regain the old numbers among them in a natural way. That applies to almost all these types of expansions. The embedding is $n \mapsto (n, 0)/_\sim$.

The transition from integers to rational numbers is analogous. Consider pairs of integers $(m, n)$, where $n \neq 0$. Heuristic: with $(m, n)$, think of $m/n$.[6]

**Definition 3.28** $(m_1, n_1) \approx (m_2, n_2) := m_1 n_2 = m_2 n_1$.

---

[6] Who is very precise should consider pairs of pairs, say $[(m, n), (k, l)]$. We make it easy for ourselves by writing the integers in their traditional form.

This is an equivalence relation (check). Define the operations $+, -, \cdot, ^{-1}$ and the relation $<$ on these pairs yourself, and show that they preserve the equivalence $\approx$. The rational numbers are equivalence classes under $\approx$. In practice, we write $m/n$ for $(m, n)/\approx$. Indicate the embedding of the integers in the rational numbers.

In fact, there is a lot to prove before we have all the basic properties of integers and rational numbers. Since it is a standard part of most courses and books on algebra or analysis, we will consider the work done.

The natural numbers form the set $\mathbb{N}$, integers form the set $\mathbb{Z}$, the rational numbers form the set $\mathbb{Q}$. The sets $\mathbb{Z}$ and $\mathbb{Q}$ are ordered in a natural way, so that $(m, n) > 0$ in $\mathbb{Z}$ if $m > n$ in $\mathbb{N}$, and $(m, n) > 0$ in $\mathbb{Q}$ if $mn > 0$ in $\mathbb{Z}$.

For constructive mathematics it is important that the basic relations are effective, in other words, that $\sim$, $\approx$ and $<$ are decidable.

Note that in effective mathematics there are broadly two meanings of "decidable". One is technical decidability from the theory of algorithms, whereby a relation $R(n, m, \dots)$ is decidable if there is an algorithm $F$ (say, a Turing Machine) that calculates the characteristic function of $R$. The other is the decidability of the intuitionistic logic: $R(n, m, \dots)$ is decidable if $R(n, m, \dots) \vee \neg R(n, m, \dots)$ is valid (has an (informal) proof). The two meanings of decidability are not independent of each other.

## 3.3  More on Orderings

At first sight, $\mathbb{N}$, $\mathbb{Z}$ and $\mathbb{Q}$ are rather different from one another. Although we can think of these such that $\mathbb{N} \subset \mathbb{Z} \subset \mathbb{Q}$, their orderings set them apart. The two that resemble each other most are $\mathbb{N}$ and $\mathbb{Z}$, because in both all points are isolated, in the sense that limits of the kind we are used to do not exist. That is not strictly true: if we keep to the usual definition of a limit, we say that $\lim_{j \to \infty} n_j = n$ if $\forall \epsilon > 0 \, \exists N \, \forall j > N (|n_j - n| < \epsilon)$. Now if we take $\epsilon = 1/2$, then from a certain $N$ onward we have $n_j = n$. In words: only sequences whose elements sooner or later become constant converge. As a consequence, the topologies of $\mathbb{N}$ and $\mathbb{Z}$ are not so exciting. With $\mathbb{Q}$ we get more interesting limits, e.g., $\lim_{n \to \infty} 1/n = 0$. On the other hand there are sequences here that do get ever closer to one another, yet have no limit. Think of rational approximations to $\sqrt{2}$.

One of the things that $\mathbb{N}$, $\mathbb{Z}$ and $\mathbb{Q}$ do have in common is that they are denumerable. In a set-theoretical sense they are therefore of the same size. A set $X$ is called *denumerable* if there is a bijection between $\mathbb{N}$ and $X$. One also says: if there is a one-one correspondence between $X$ and $\mathbb{N}$.

Here is a bijection from $\mathbb{N}$ to $\mathbb{Z}$:

$$f(2n) = n, \ f(2n + 1) = -n - 1.$$

A bijection from $\mathbb{N}$ to $\mathbb{Q}$ is a bit more complicated. Here are three ideas we will use to construct one.

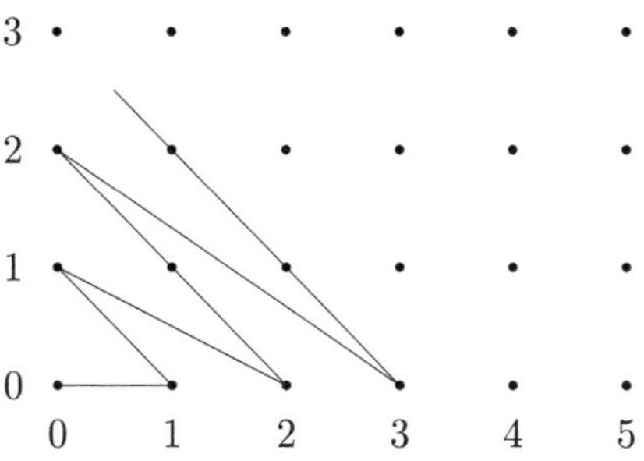

**Fig. 3.1** Beginning a walk through all points of the grid $\mathbb{N} \times \mathbb{N}$

(1) The set of points $(n, m)$ on the grid $\mathbb{N} \times \mathbb{N}$ is denumerable. This is illustrated by the walk in Fig. 3.1. So we have a bijection $f : \mathbb{N} \to \mathbb{N} \times \mathbb{N}$.

(2) Let $X$ be a family of denumerably many disjoint denumerable sets, each indexed by a natural number $i$, with $X_i = \{x_{i,0}, x_{i,1}, x_{i,2}, \ldots\}$. Define $g$ by $g(i, j) = x_{i,j}$, then $g$ is a bijection from $\mathbb{N} \times \mathbb{N}$ to $\bigcup_i X_i = X$. Now $f$ followed by $g$ is a bijection from $\mathbb{N}$ to $X$.

(3) Construct an enumeration (without repetition) of the elements of $\mathbb{Q}$ in the interval $[0, 1)$:

$$0, \frac{1}{2}, \frac{1}{3}, \frac{2}{3}, \frac{1}{4}, \frac{3}{4}, \frac{1}{5}, \frac{2}{5}, \frac{3}{5}, \frac{4}{5}, \frac{1}{6}, \ldots$$

Do the same for each interval $[i, i + 1)$. Thus, we obtain disjoint denumerable sets $\mathbb{Q}_i$ with $\bigcup_i \mathbb{Q}_i = \mathbb{Q}$.

Combine ideas (1) to (3) to construct a bijection $j$ from $\mathbb{N}$ to $\mathbb{Q}$.

The natural ordering of $\mathbb{Q}$ is highly interesting. It is a dense ordering without endpoints, where "dense" means that in between any two elements there is a third element. The surprising thing is that these two properties completely determine the denumerable dense ordering without endpoints: any two sets with a denumerable dense ordering without endpoints are isomorphic. The proof goes back to Georg Cantor, and the proof is too amusing not to mention. It is known as the "zig-zag"-method.

Consider two such sets, say $X = \{x_0, x_1, x_2, \ldots\}$ and $Y = \{y_0, y_1, y_2, \ldots\}$. The isomorphism we are looking for is now defined step by step, where we keep turning from $X$ to $Y$ and back again.

(1) Define $f(x_0) = y_0$.

(2) Take $x_1$ and check whether $x_0 < x_1$ or rather $x_1 < x_0$. Let us say that $x_0 < x_1$; now look for the first index $i$ such that $y_0 < y_i$ and define $f(x_1) = y_i$.

(3) Now move to $y_1$. There are two cases:

    (a) $y_1 \neq f(x_1)$. Determine where $y_1$ lies with respect to $f(x_0)$ and $f(x_1)$. There are $3 \times 2 \times 1$ possibilities:

$$
\begin{aligned}
y_1 \;&<\; f(x_0) \;<\; f(x_1) \\
y_1 \;&<\; f(x_1) \;<\; f(x_0) \\
f(x_0) \;&<\; y_1 \;<\; f(x_1) \\
f(x_0) \;&<\; f(x_1) \;<\; y_1 \\
f(x_1) \;&<\; y_1 \;<\; f(x_0) \\
f(x_1) \;&<\; f(x_1) \;<\; y_0
\end{aligned}
$$

Because of the dense ordering and the absence of endpoints, we can find in $X$ the first $x_j$ with the same position with respect to $x_0$ and $x_1$ as $y_1$. Now define $f(x_j) = y_1$.

(b) $y_1 = f(x_1)$. Move on to $y_2$ and repeat the argument.

Thus, we keep choosing the first element in $X$ or $Y$ that has not been looked at yet. Finish the argument yourself.

# Chapter 4
# Real Numbers

**Abstract** We construct real numbers via Cauchy sequences, utilizing the Axiom of Countable Choice, which can be justified by appealing to the Proof Interpretation of logic and the construction of the natural numbers. Equality of real numbers is generally undecidable, so we look for positive concepts. The ordering a < b is defined by a positive lower bound on distance; the classical law of trichotomy then fails. The positive relation of apartness comes to replace inequality. The classical topological concepts like supremum and infimum are shown to be applicable only for sets satisfying the constructive property of being "located".

## 4.1 Going Beyond the Rational Numbers

The transition from rational numbers to real numbers is a critical step; here we move from finitely specifiable objects to essentially infinite objects.

Real numbers allow us to treat continua mathematically.

Originally, the continuum is directly given to us in perception or intuition. Opinions differ whether the fundamental continuum is the spatial one, the temporal one, or both. For Brouwer, it was the temporal one, in a specific sense:

> Of course we mean here *intuitive time*, which must be distinguished from *scientific time*. By means of experience and very much a posteriori it appears that scientific time can suitably be introduced for the cataloguing of phenomena, as a one-dimensional coordinate having a one-parameter group. [28, p. 61]

Thus, the continuum he is talking about is not that of physical time, but of the time whose passing is intrinsic to consciousness, to our lived awareness.[1] He also emphasized, as various philosophers had done before, that the  continuous cannot be constructed out of the discrete, and the  discrete not out of the continuous. They are equally fundamental, and mutually dependent: the one can only be grasped in its contrast to the other.

---

[1] This has been analyzed extensively in so-called phenomenology, the philosophical school founded by Edmund Husserl. It arguably makes for a better fit than Kant's ideas about mathematics and time, from which  Brouwer started in 1907, but seems to have dropped around the time he came up with choice sequences. Here we will leave it at that. (But see also the remark at the end of Sect. 6.4.)



D. van Dalen et al., *Intuitionistic Analysis*, Springer Undergraduate Mathematics Series, https://doi.org/10.1007/978-3-032-16491-9_4

Now it is very nice to have a continuum in (your) mind, but you can't do a lot of mathematics with that yet. We need to be able to relate numbers to it in some way that is characteristic for the continuum, numbers that we can add, compare, and so on. It is not that hard to place the integers on the continuum. Then we can enumerate the rational numbers, e.g., $1/2, 1/3, 2/3, 1/4, 3/4, \ldots$ and place them on the continuum, too. Now $\mathbb{Q}$ forms a nice dense set on the continuum, but there are still gaps in it, so it cannot model continuity. We have already seen that $\sqrt{2}$ is irrational, so if we go through $\mathbb{Q}$, we will not come across $\sqrt{2}$. To fill those gaps we add at one blow all real numbers as limits of convergent sequences. Surprisingly, this works, but it does require some effort.

We are now going to carry out this process of extension in some detail. The method dates back to the nineteenth century, where reducing the continuum to arithmetic was called *arithmetization*. All standard textbooks present this (or a similar) method, but we must pay attention to it, because in intuitionistic mathematics non-obvious refinements can arise.

## 4.2  The Axiom of Countable Choice

In set theory, Ernst Zermelo's Axiom of Choice (AC) says that "if a family of non-empty sets $X_i$ is given, then there is a function from that family to the union of those sets that from each $X_i$ chooses an element $x$". There exist various specialized forms and equivalents of it. These are very frequently used in classical mathematics. When trying to find constructive analogs of these uses, one immediately asks where the construction method for such $x$ is; it then turns out that, constructively, most versions of AC do not hold![2] We will return to this in Sect. 6.8.

Why bring up AC here? Because there is a specific form of AC, the "Axiom of Countable Choice", that

(1) is, after all, constructively correct, *as a matter of the Proof Interpretation and the nature of the natural numbers*. So it can be justified from what we have seen in the previous two chapters.
(2) is used in most constructive theories of real numbers and functions, the topics of this chapter and the next.[3]

The Axiom of Countable Choice states

$$\forall n \in \mathbb{N} \; \exists m \in \mathbb{N} \; A(n, m) \rightarrow \exists f \in \mathbb{N}^{\mathbb{N}} \; \forall n \; A(n, f(n)) . \tag{4.1}$$

---

[2] For a detailed discussion, see [60]. In fact, an animated discussion about the validity of the axiom arose immediately after its introduction, even before such things as weak counterexamples were known; see [44].

[3] It is interesting to see what constructive mathematics looks like without the Axiom of Countable Choice. See for example [73].

Notation: AC-NN or $AC_{00}$, as $f$ is a function from $\mathbb{N}$ to $\mathbb{N}$, or from objects of the lowest type to objects of the lowest type. That this is a special version of AC is seen by associating to each $n \in \mathbb{N}$ the set of $m \in \mathbb{N}$ such that $A(n, m)$ holds, and then considering the $n$ as indices to such sets, thereby obtaining a family $X_n$. From this perspective, the antecedent of AC-NN means that not only the sets in this family are not empty, but, positively, that of each we can construct an element.

An informal justification of AC-NN runs as follows. Assume that the antecedent is true, i.e., that we have a proof of it. So we have a method that given any $n \in \mathbb{N}$ yields a proof of $\exists m \in \mathbb{N} \; A(n, m)$, that is, a value $m = g(n)$ for an appropriate construction method $g$, together with a construction $g'(n)$ that proves $A(n, m)$. Combining these methods, we can say: if we have a proof of the antecedent, then we have a method to construct, for each $n \in \mathbb{N}$, an $m$ and a proof of $A(n, m)$; that method is an $f$ whose existence the consequent claims.[4]

Clearly, for this justification the range of the choice function (the second N in AC-NN) does not really matter, so it yields AC-N ($AC_0$):

$$\forall n \in \mathbb{N} \; \exists d \in D \; A(n, d) \rightarrow \exists f \in D^{\mathbb{N}} \; \forall n \; A(n, f(n)) \tag{4.2}$$

for any $D$, not just $\mathbb{N}$.

Somewhat more formally, the argument for AC-NN (and AC-N) looks like this.

Let us assume we have a proof $a$ of the antecedent, so $a : \forall n \in \mathbb{N} \; \exists m \in \mathbb{N} \; A(n, m)$. By the definition of "proof" (Sect. 2.5), this means that $a$ is a construction (method) that yields for every $n$ a proof $a(n) : \exists m \, A(n, m)$. Now applying the clause for $\exists$ in that definition we get: the method $a$ yields for every $n$ a proof $p_1(a(n)) : A(n, p_0(a(n)))$. Here $p_0$ and $p_1$ are projection operators: constructions that, when applied to a pair, yield the first and the second element, respectively. There is also the operator $p$ which, conversely, takes two objects and forms their ordered pair.

A proof $c$ for the consequent would consist in a pair $(c_0, c_1)$ where $c_0$ is a function from $\mathbb{N}$ to $\mathbb{N}$ and $c_1 : \forall n \, A(n, c_0(n))$. Using what our assumed proof of the antecedent gives us, we can indeed obtain such a $c$, by setting $c_0 = \lambda n.p_0(a(n))$ and $c_1 = \lambda n.p_1(a(n))$. The device we just used here is that of $\lambda$-abstraction: roughly, it enables us to turn a description of a method for doing things into a function, the function that applies that method to its argument. In this case, $c_0$ is the function that is the "witness" for the existential quantification over functions in the consequent.

We have proceeded by operating on an assumed proof of the antecedent of (4.2). To prove that implication, we need to turn this way of proceeding into a mapping $h$ from proofs of the antecedent to proofs of the consequent. This is again achieved by $\lambda$-abstraction:

$$h := \lambda a.p(\lambda n.p_0(a(n)), \lambda n.p_1(a(n)))$$

---

[4] For Brouwer, a method for associating to each $n \in \mathbb{N}$ an $m \in \mathbb{N}$ need not even consist in an algorithm or law; he accepted functions that "become freely", in this case choice sequences $\mathbb{N} \rightarrow \mathbb{N}$. We will come back to choice sequences in Chap. 6.

It is important to see that the construction of the $f$ that witnesses the consequent of the implication may depend on a proof of the antecedent. Specifically, this is shown by the occurrence of the proof $a$ in the definition of $c_0$. Thus, the role of the Proof Interpretation in the justification of AC-NN is clear. The reader will have noticed that, in contrast, we have not said anything about the dependence we heralded above of the constructibility of this function on the fact that its domain is that of the natural numbers. That doesn't have much to do with the construction method itself, but with the reason for calling it a function. If that sounds mysterious, skip to Sect. 6.8.

## 4.3  Cauchy Sequences

**Definition 4.1**  A sequence $(a_n)$ of rational numbers is called a *Cauchy sequence* if for every $k$ there is an $n$ with the property $|a_n - a_m| < 2^{-k}$ for all $m > n$. Formally, this reads:

$$\forall k \exists n \forall m (|a_n - a_{n+m}| < 2^{-k})$$

Note that the $\forall k (\ldots 2^{-k})$ here represents the idea of the "arbitrarily small", usually associated with $\epsilon$. You could also say instead "for every positive rational number $\epsilon$, there is ...". In intuitionistic mathematics we stick to $2^{-k}$ because then its calculation is kept at the most fundamental level, i.e., that of the natural numbers. It also helps to make calculations easier and clearer later on.

The core of the definition is in the *existence*; the $n$ mentioned must really be constructible.

We illustrate this point with a weak counterexample.

Consider the following sequence: as long as every even number larger than 2 and smaller than $n$ is the sum of 2 primes, we set $a_n = 1$; once an even number $n$ has occurred that is not the sum of two prime numbers, we set $a_n = 0$. As the Goldbach Conjecture is still open, we don't know whether the sequence $a_n$ is constant or makes the jump to 0. For the classical mathematician this $(a_n)$ is a proper Cauchy sequence, since the Goldbach Conjecture is correct or incorrect. If it is correct, then $(a_n)$ is constant 1, which is evidently a Cauchy sequence. If it is incorrect, then $(a_n)$ eventually becomes constant 0, and is again evidently a Cauchy sequence.

The constructivist, however, runs into a difficulty. Look at the definition of Cauchy sequence, and consider the case $k = 1$. We now have to calculate an $n$ so that all values after $a_n$ are at a distance less than $1/2$ from it. But whoever can calculate that $n$ has automatically solved the conjecture. So from a constructive standpoint we cannot say that $(a_n)$ is a Cauchy sequence. Maybe 100 years from now we will be able to do so, but that doesn't do us much good now.

Conversely, whenever we *do* have a proof that some sequence is a Cauchy sequence, this means we have a proof of a proposition of the form $\forall k \exists n (\ldots)$, and then AC-NN makes explicit that this proof involves a function that allows us to calculate or to determine, from a given $k$, a suitable $n$. This also holds when we are speaking hypothetically: when we *assume* that we have a proof of $\forall k \exists n(\ldots)$, we thereby also *assume* that there exists a certain function.

## 4.4  From Cauchy Sequences to Real Numbers

The role of Cauchy sequences is, in popular terms, to function as numbers of a kind that can be used in a mathematical model of the continuum: real numbers. Real numbers generally do not correspond to already constructed rational numbers. On the other hand, the rational numbers are in a natural way included in the reals; we will return to the matter at the end of Sect. 4.5.

There is another point that deserves attention: different Cauchy sequences can represent the same real number! Take $(a_n)$ and $(b_n)$ with $a_n = 0$ for all $n$, and $b_n = 2^{-n}$ for all $n$. The solution to this problem begins with the following definition.

**Definition 4.2**  The sequences $(a_n)$ and $(b_n)$ are *equivalent*, notation $(a_n) \sim (b_n)$, if $\forall k \exists n \forall m (|a_{n+m} - b_{n+m}| < 2^{-k})$.

In words: for every $k$ there is an index from which the corresponding values differ by less than $2^{-k}$.

**Lemma 4.3**  $\sim$ *is an equivalence relation.*

***Proof***  $\sim$ has each of the three defining properties of an equivalence relation:

(1)  $\sim$ is reflexive, i.e., $(a_n) \sim (a_n)$. Trivial.
(2)  $\sim$ is symmetric. Also trivial.
(3)  $\sim$ is transitive. Here is something to calculate. Determine

$$
\begin{aligned}
n_1 \quad &\text{such that } \forall m (|a_{n_1+m} - b_{n_1+m}| < 2^{-k-1})\,, \\
n_2 \quad &\phantom{\text{such that }} \forall m (|b_{n_2+m} - c_{n_2+m}| < 2^{-k-1})\,.
\end{aligned}
$$

Let $n = \max(n_1, n_2)$, so that for this $n$ the two inequalities hold at the same time. Now for all $m$ it holds that

$$
|a_{n+m} - c_{n+m}| \le |a_{n+m} - b_{n+m}| + |b_{n+m} - c_{n+m}| < 2^{k-1} + 2^{k-1} = 2^{-k}
$$

Thus one finds a suitable $n$ for the given $k$, i.e., $(a_n) \sim (c_n)$.

$\square$

A practical hint for readers who like to check things for themselves. Definitions are given in a special form. It is often required that a particular term be below a limit, say $2^{-k}$. When we start work on a proof, we often can't see how small to choose certain other terms. The result is that the estimate may be too large, for example $< 2^{-k+3}$. Then we can always see to it, by adjusting (decreasing) the initial limits, that the final estimate is indeed $< 2^{-k}$. In other words, whoever has come so far as to have determined a bound can consider his task (virtually) finished.

With the definition of equivalence we could perhaps have asked a little more, namely that after $n$ all the following elements come arbitrarily close to each other. That turns out to be true:

**Lemma 4.4** *If* $(a_n) \sim (b_n)$, *then* $\forall k \exists n \forall p \forall q (|a_{n+p} - b_{n+q}| < 2^{-k})$

***Proof*** Because $(a_n)$ and $(b_n)$ are Cauchy sequences and $(a_n) \sim (b_n)$ we can, for each $k$, determine

$$
\begin{aligned}
n_1 \quad &\text{such that} \quad \forall m(|a_{n_1} - a_{n_1+m}| < 2^{-k-2}), \\
n_2 \quad &\phantom{\text{such that}} \quad \forall m(|b_{n_2} - b_{n_2+m}| < 2^{-k-2}), \\
n_3 \quad &\phantom{\text{such that}} \quad \forall m(|a_{n_3+m} - b_{n_3+m}| < 2^{-k-2}).
\end{aligned}
$$

Now choose $n = \max(n_1, n_2, n_3)$. Then for every $p$ and $q$,

$$
\begin{aligned}
|a_n - a_{n+p}| &< 2^{-k-2}, \\
|b_n - b_{n+q}| &< 2^{-k-2}, \\
|a_n - b_n| &< 2^{-k-2}.
\end{aligned}
$$

From this it follows that

$$
\begin{aligned}
|a_{n+p} - b_{n+q}| &\leq |a_{n+p} - a_n| + |a_n - b_n| + |b_n - b_{n+q}|, \\
&< 3 \cdot 2^{-k-2}, \\
&< 4 \cdot 2^{-k-2} = 2^{-k}.
\end{aligned}
$$

$\square$

**Exercise 4.5**

(1) If $(a_n)$ is a Cauchy sequence and $(a_n) \sim (b_n)$, then $(b_n)$ is also a Cauchy sequence.
(2) Let $(a_n)$ be a Cauchy sequence $b_n = a_{n+1}$, then $(a_n) \sim (b_n)$.
(3) Ditto for $b_n = a_{2n}$.
(4) Ditto for $b_n = a_{n^2}$.
(5) If $(b_n)$ is an infinite subsequence of the Cauchy sequence $(a_n)$ then $(a_n) \sim (b_n)$.
(6) Let $(a_n)$ be a Cauchy sequence; show that $(a_n) \sim (a_n + 2^{-n})$
(7) Define, given two Cauchy sequences $(a_n)$ and $(b_n)$, the sequence $(c_n)$ by

$$
\begin{aligned}
c_{2n} &= a_n \\
c_{2n+1} &= b_n
\end{aligned}
$$

Show: $(c_n)$ is a Cauchy sequence $\leftrightarrow (a_n) \sim (b_n)$.

(8) Show that for every Cauchy sequence $(a_n)$ there exists an equivalent sequence $(b_n)$ with $\forall k \forall m (|b_k - b_{k+m}| < 2^{-k}$. The convergence rate of $(b_n)$ can be arranged in this way.

Now that we have the equivalence relation, we can identify equivalent Cauchy sequences.

**Definition 4.6** The equivalence classes of Cauchy sequences under the relation $\sim$ are called *real numbers*. The real number determined by $(a_n)$ is $(a_n)/_\sim$. We say that $(a_n)$ is a *representative* of $(a_n)/_\sim$.

Every $(b_n)$ with $(b_n) \sim (a_n)$ is of course also a representative. Therefore, we can define equality of real numbers as equivalence of respective representatives:

**Definition 4.7** Let $a$ be a real number given by $(a_n)$ and $b$ a real number given by $(b_n)$. $a = b := (a_n) \sim (b_n)$.

We denote the set of real numbers by $\mathbb{R}$.

When we work with real numbers, we often fall back on their representatives; of course we don't want random properties of the chosen representative to rule (for example, if in the chosen sequence every even $a_{2n}$ equals its successor $a_{2n+1}$, then that is totally irrelevant to the real number being determined). We must therefore show for each new concept that it *is independent of the choice of the representative*.

More precisely, if a property $A(a)$, respectively a relation $B(a, b)$ holds for real numbers $a = (a_n)/_\sim$, $b = (b_n)/_\sim$, then we have to show that $(a_n) \sim (a'_n)$, $(b_n) \sim (b'_n) \rightarrow A(a')$ and $B(a', b')$, where $a' = (a'_n)/_\sim$, $b' = (b'_n)/_\sim$. After all $a = a'$ and $b = b'$.

Now that the real numbers and equality on them have been introduced, we need to define basic relations and operations, such as

| | |
|---|---|
| addition | $a + b$ |
| apartness | $a \# b$ |
| inverse | $a^{-1}$ |
| multiplication | $a \cdot b$ or $ab$ |
| opposite | $-a$ |
| order | $a < b$ or $b > a$ |

Order is taken on in the next section, the rest will come after that.

## 4.5  Ordering the Real Numbers

We agree to use the following notation: $a = (a_n)/_\sim$, $a' = (a'_n)/_\sim$, $b = (b_n)/_\sim$, etc.

**Definition 4.8**  $a < b := \exists k \exists n \forall m (b_{n+m} - a_{n+m} > 2^{-k})$

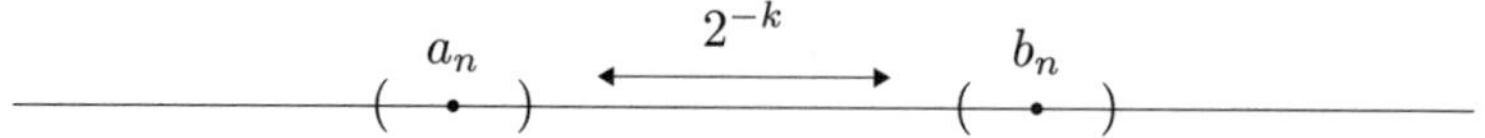

Fig. 4.1 When one real number is greater than another

Fig. 4.2 The choice of representative does not change the order

See Fig. 4.1. In words: from a certain index $n$ on the elements of $(b_n)$ are at least $2^{-k}$ greater than the corresponding elements of $(a_n)$ for a fixed $k$. So the ordering $<$ indicates a positive property, given by the presence of a certain positive rational minimum distance.[5]

**Lemma 4.9** *The ordering is independent of the choice of the representative: if $a \sim a', b \sim b'$ and $a < b$, then $a' < b'$.*

*Advice*: these kinds of proofs are not the most interesting part of mathematics, but they are necessary for exactness. Once you understand the principle, you can prove all similar claims. You can skip the proofs now, and do them later when there is nothing else to do. In no time you will see how easy it is.

The idea behind the proof will be this (Fig. 4.2). From a certain $n$ on, the $a_n$ lie well to the left of the $b_n$. Make sure that all $a'_m$ and $b'_m$ from a certain $n$ on are sufficiently close to such $a_n$ and $b_n$, and then show that the $a'_n$ are well to the left of the $b'_n$.

**Proof** We determine a

$$
\begin{aligned}
k, n_1 \quad &\text{such that for all} \quad m > n_1, \quad b_m - a_m > 2^{-k}, \\
n_2 \quad & m > n_2, \quad |a_m - a'_m| < 2^{-k-3}, \\
n_3 \quad & m > n_3, \quad |b_m - b'_m| < 2^{-k-3}, \\
n_4 \quad & m, p > n_4, \quad |a_m - a_p| < 2^{-k-3}, \\
n_5 \quad & m, p > n_5, \quad |b_m - b_p| < 2^{-k-3}.
\end{aligned}
$$

Then we set $n = \max(n_1, n_2, n_3, n_4, n_5) + 1$. Now consider $a'_m, b'_m$ for $m > n$, then

$$
|a_n - a'_m| \le |a_n - a_m| + |a_m - a'_m| < 2^{-k-2},
$$
$$
|b_n - b'_m| \le |b_n - b_m| + |b_m - b'_m| < 2^{-k-2}.
$$

Now we will show that $a'_m$ is a bit to the left of $b'_m$. Suppose $|b'_m - a'_m| \le 2^{-k-1}$. Then

---

[5] Brouwer wrote $a \lozenge b$ for this, and used $a < b$ for $\neg\neg(a \lozenge b)$.

$$|b_n - a_n| \leq |b_n - b'_m| + |b'_m - a'_m| + |a'_m - a_n|,$$
$$< 2^{-k-2} + 2^{-k-1} + 2^{-k-2},$$
$$= 2^{-k}.$$

But since $n > n_1$, this contradicts our condition on $n_1$.

So $b'_m - a'_m > 2^{-k-1}$. Therefore from the index $n$ on, all $a'_m$ are at least $2^{-k-1}$ to the left of $b'_m$. $\qquad\square$

We proved $b'_m - a'_m > 2^{-k-1}$ indirectly (see Sect. 2.4)—haven't we cheated? Not really, because for rational numbers the order relations are decidable. We can always determine whether $b'_m - a'_m \leq 2^{-k-1}$ or $b'_m - a'_m > 2^{-k-1}$. Whenever we can exclude the one, we know that the other will hold.

When dealing with this kind of argument for the first time, one may be tempted to think that since what is requested is a $k$ such that $b'_m - a'_m > 2^{-k}$, coming up with a $k$ such that $b'_m - a'_m > 2^{-k-1}$, i.e., $> 2^{-(k+1)}$, would not be good enough. But what is requested is *some* appropriate $k$, so we simply let the $k+1$ we found be our new $k$, and give that as our answer.

**Exercise 4.10** Show $a < b \leftrightarrow \exists k \exists n \forall m_1 \forall m_2 (b_{n+m_1} - a_{n+m_2} > 2^{-k})$.

Once we have defined the ordering for real numbers, we can ask ourselves how the rational numbers fit into the picture. For the ordering, this can be seen from the following theorem:

**Theorem 4.11** $a < b \to \exists r \in \mathbb{Q}\ (a < r < b)$, *i.e., the rational numbers are dense in the real numbers.*

***Proof*** Let $a = (a_n)$, $b = (b_n)$. By Exercise 4.10,

(1) $\exists k \exists n_1 \forall m_1 \forall m_2 (b_{n_1+m_1} - a_{n_1+m_2}) > 2^{-k}$

We also know

(2) $\exists n_2 \forall m_1 \forall m_2 (|a_{n_2+m_1} - a_{n_2+m_2}| < 2^{-k-2})$
(3) $\exists n_3 \forall m_1 \forall m_2 (|b_{n_3+m_1} - b_{n_3+m_2}| < 2^{-k-2})$

Choose $n = \max(n_1, n_2, n_3)$. Now consider any sufficiently distant elements $b_{n+m_1}$ and $b_{n+m_2} + a_{n+m_2}$ of the Cauchy sequences $b$ and $(b+a)/2$, respectively. Using (1) and (3) we find

$$b_{n+m_1} - \frac{b_{n+m_2} + a_{n+m_2}}{2} = \frac{b_{n+m_1} - b_{n+m_2}}{2} + \frac{b_{n+m_1} - a_{n+m_2}}{2} > 2^{-k-1}.$$

So the rational number $(b_{n+m_2} + a_{n+m_2})/2$ is smaller than $b$. Similarly, we see that it is greater than $a$. $\qquad\square$

$$a_{n+m} \qquad\qquad b_{n+m} \qquad\qquad\qquad c_{n+m}$$

**Fig. 4.3**  Transitivity of the order

We see that the ordering relation has been defined correctly. Its most important properties are listed below.

**Theorem 4.12**

(1)  $a = b \wedge a < c \rightarrow b < c$
(2)  $b = c \wedge a < c \rightarrow a < b$
(3)  $a < b \wedge b < c \rightarrow a < c$
(4)  $a < b \rightarrow \neg a = b$
(5)  $a < b \rightarrow \neg b < a$
(6)  $a = b \rightarrow \neg a < b$

*Proof*

(1)  $a = b \leftrightarrow (a_n) \sim (b_n)$. If we now replace in $(a_n)/_\sim < (c_n)/_\sim$ the sequence $(a_n)$ by the equivalent sequence $(b_n)$, it follows from Lemma 4.9 that $(b_n)/_\sim < (c_n)/_\sim$, so $b < c$.
(2)  Ditto.
(3)  See Fig. 4.3. Determine

$$k_1, n_1 \;\; \text{such that}\;\; \forall m\left(|b_{n_1+m} - a_{n_1+m}| > 2^{-k_1}\right),$$
$$k_2, n_2 \qquad\qquad \forall m\left(|c_{n_2+m} - b_{n_2+m}| > 2^{-k_2}\right).$$

Let $k = \max(k_1, k_2)$ and $n = \max(n_1, n_2)$, then
$$\forall m\left((b_{n+m} - a_{n+m} > 2^{-k}) \wedge (c_{n+m} - b_{n+m} > 2^{-k})\right),$$
and so

$$\begin{aligned}
(c_{n+m} - a_{n+m}) &= (c_{n+m} - b_{n+m}) + (b_{n+m} - a_{n+m}), \\
&> 2^{-k} + 2^{-k}, \\
&= 2^{-k+1}.
\end{aligned}$$

Ergo $a < c$.

$\square$

**Exercise 4.13**  Prove (4)–(6) yourself.

In classical mathematics the ordering satisfies an important property, the so-called principle of *trichotomy*: $a < b \vee a = b \vee b < a$.

Intuitively, trichotomy holds for rational numbers, because we can simply test which of the three is the case. Real numbers require an infinite test, which in general cannot be performed, leaving the order relations undecidable. This allows us to

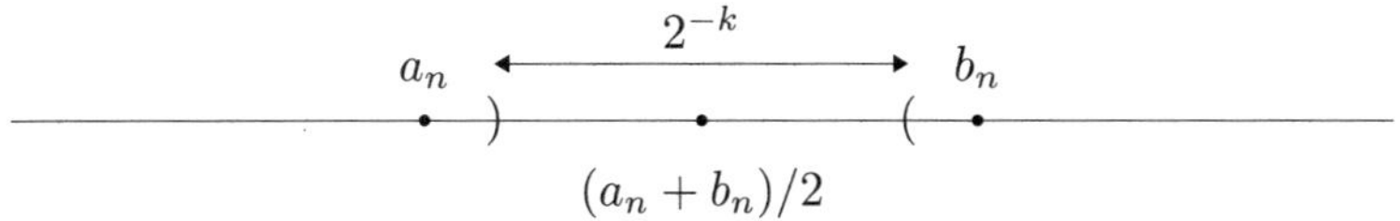


**Fig. 4.4** The midpoint between certain corresponding intervals in Cauchy sequences representing real numbers one of which is greater than the other

devise a weak counterexample; it is actually the one we saw in Chap. 1, now presented somewhat differently. For convenience, we now introduce an abbreviation: $N_{20}(n) :=$ "the $n$th decimal after the point of $\pi$ is preceded by a sequence of 20 nines". We define a Cauchy sequence $(a_n)$ as follows: write successively the decimals of $\pi$ and at the same time the elements of $(a_n)$. The sequence $(a_n)$ follows the sequence $(2^{-n})$ until 20 nines are passed, after which the elements in the sequence will remain constant.

Expressed algebraically:

$$a_n = \begin{cases} 2^{-n} & \text{if } \forall p < n \neg N_{20}(p) \\ 2^{-p} & \text{if } p < n \wedge N_{20}(p) \wedge (\forall m < p)\neg N_{20}(m) \end{cases}$$

Note that $(a_n) \sim (0)$ if $\pi$ does not contain a sequence of 20 nines and that $a = (a_n)/_\sim > 0$ if $(a_n)$ eventually becomes constant; $0 > (a_n)/_\sim$ is excluded. We now see that $a > 0 \vee a = 0$ is equivalent to "either there is a sequence of 20 nines in $\pi$, or there is not". There is no evidence whatsoever for this last statement, even assuming trichotomy.

In this example we do know something: $a$ cannot be negative. We can make the indeterminacy worse by making the sequence oscillate:

$$a_n = \begin{cases} (-2)^{-n} & \text{if } \forall p < n \neg N_{20}(p) \\ (-2)^{-p} & \text{if } p < n \wedge N_{20}(p) \wedge (\forall m < p)\neg N_{20}(m) \end{cases}$$

In this case, the position relative to 0 is completely undetermined: we do not know $a < 0 \vee a = 0 \vee a > 0$. This also means that we cannot yet have a method to convert the Cauchy sequence that defines $a$ to a decimal expansion. In other words, we have a weak counterexample to the thesis that every real number has a decimal expansion.[6]

Instead of trichotomy, we can prove the following useful proposition.

**Theorem 4.14**  $a < b \to a < c \vee c < b$

***Proof*** See Fig. 4.4. Determine a $k$ and an $n_1$ such that $\forall m \, (b_{n_1+m} - a_{n_1+m} > 2^{-k})$, and further choose

---

[6] Indeed, in 1920 Brouwer gave a talk, and the next year published a paper, with the (translated) title "Does every real number have a decimal expansion?" [22].

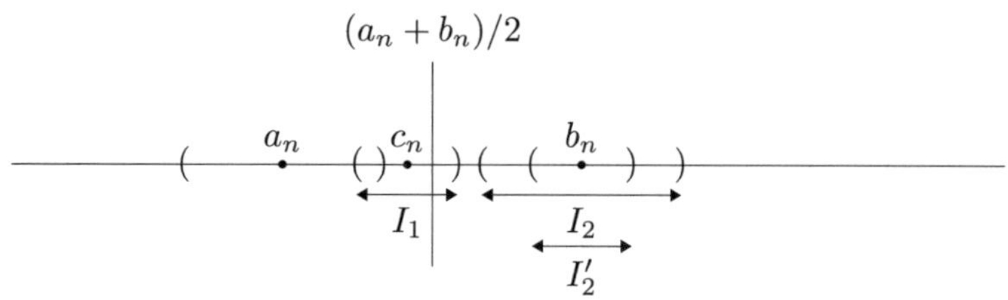

**Fig. 4.5** The intervals around $a_n$ and $c_n$ may overlap (and here they do), but not those around $c_n$ and $b_n$

$$
\begin{aligned}
n_2 \quad \text{such that} \quad & |a_{n_2} - a_{n_2+m}| < 2^{-k-3}, \\
n_3 \qquad\qquad\quad & |b_{n_3} - b_{n_3+m}| < 2^{-k-3}, \\
n_4 \qquad\qquad\quad & |c_{n_4} - c_{n_4+m}| < 2^{-k-3}.
\end{aligned}
$$

Let $n = \max(n_1, n_2, n_3, n_4)$. We are now going to look at the location of $c_n$. This number falls into either of the intervals $(-\infty, (a_n + b_n)/2]$, $[(a_n + b_n)/2, \infty)$. Consider the first case. The intervals $I_1 = (c_n - 2^{-k-3}, c_n + 2^{-k-3})$ and $I_2 = (b_n - 2^{-k-2}, b_n + 2^{-k-2})$ cannot intersect (Fig. 4.5). For, suppose the interval $I_1$ overlaps $I_2$, then $b_n - c_n < 2^{-k-1}$. Since by hypothesis $c_n \leq (a_n + b_n)/2 < b_n$, now also $b_n - (a_n + b_n)/2 < 2^{-k-1}$, hence $b_n - a_n < 2^{-k}$. Contradiction.

Since $I_1$ is to the left of $I_2$, and on the right has endpoint at most $(a_n + b_n)/2 + 2^{-k-3}$, the distance between $I_1$ and $I_2' = (b_n - 2^{-k-3}, b_n + 2^{-k-3}) \geq 2^{-k-3} + 2^{-k-3} = 2^{-k-2}$, and so $b_{n+m} - c_{n+m} > 2^{-k-2}$ for all $m$. That is, $c < b$.

The second case is analogous, yielding $a < c$. $\qquad\qquad\qquad\qquad\qquad\square$

**Definition 4.15**

$$
\begin{aligned}
a > b &:= b < a \\
a \geq b &:= \neg a < b \\
a \leq b &:= \neg b < a
\end{aligned}
$$

A word of warning. We are used to pronouncing $\geq$ as "greater than or equal to", because that is the usual meaning given to the sign. That is *not* the interpretation we use here; for us, it is the negation of a positive property, hence a negative property. We should say "not smaller than". One can stick to the old habit, as long as one keeps in mind what is really meant. Note also that negative properties can be proved by contradiction; see the proof of Lemma 4.28(3) for an example.

**Definition 4.16** The *open interval* $(a, b)$ is the set $\{x \in \mathbb{R} \mid a < x < b\}$, where $a < b$. The *closed interval* $[a, b]$ is the set $\{x \in \mathbb{R} \mid a \leq x \leq b\}$, where $a \leq b$.

For convenience, we allow the notations $(-\infty, a)$ and $(a, \infty)$, where $(-\infty, a) = \{x \in \mathbb{R} \mid x < a\}$, etc.

Closed intervals are closed under limits, that is, if $(c_n)$ converges and $\forall n (c_n \in [a, b])$, then $\lim(c_n) \in [a, b]$. This is a consequence of

**Theorem 4.17** $\forall n (c_n \leq d_n) \to \lim(c_n) \leq \lim(d_n)$

**Exercise 4.18** Prove this theorem.

Note that we can also define closed intervals for pairs $a, b$ where the location of $a$ and $b$ relative to each other is unknown. We simply set $[a, b] = [\min(a, b), \max(a, b)]$. The exact definitions of min and max on real numbers will be given in Definition 4.36.

We now return to the question raised at the beginning of Sect. 4.4: are rational numbers also real numbers? Strictly speaking: no. Real numbers are equivalence classes of Cauchy sequences, and rational numbers are elements of Cauchy sequences. Totally different things. But we can consider the Cauchy sequence that converges to the rational number $r$ simply because each of its elements is $r$ itself. This suggests the following

**Definition 4.19** For $r \in \mathbb{Q}$, let $\langle r \rangle$ denote the constant sequence $(r_n)$ with $r_n = r$ for all $n$. Then $\langle r \rangle / {\sim}$ is the embedding of $r$ into $\mathbb{R}$.

It is clear that this embedding preserves everything that matters, e.g., $r < s \to \langle r \rangle / {\sim} < \langle s \rangle / {\sim}$, etc.

We will immediately indulge in some sloppiness: we also denote the embedded rational number $\langle r \rangle$ with $r$. Misunderstanding is almost impossible.

## 4.6  Operations and the Apartness Relation

We introduce the most important operations without comment. The standard technique is the reduction to a termwise operation.

**Definition 4.20**

$$(a_n) + (b_n) := (a_n + b_n)$$
$$(a_n) \cdot (b_n) := (a_n b_n)$$
$$-(a_n) := (-a_n)$$

**Lemma 4.21**

*(1)* $(a_n) \sim (a_n') \wedge (b_n) \sim (b_n') \to (a_n + b_n) \sim (a_n' + b_n')$
*(2)* $(a_n) \sim (a_n') \wedge (b_n) \sim (b_n') \to (a_n b_n) \sim (a_n' b_n')$
*(3)* $(a_n) \sim (a_n') \to (-a_n) \sim (-a_n')$

We will prove (2), using the following minor

**Lemma 4.22** *Every Cauchy sequence is bounded, i.e., there is a natural number $p$ such that $\forall n (|a_n| < p)$.*

**Proof of Lemma** 4.21 (2)

$$
\begin{aligned}
|a_n b_n - a_n' b_n'| &= |a_n b_n - a_n b_n' + a_n b_n' - a_n' b_n'| \\
&\leq |a_n b_n - a_n b_n'| + |a_n b_n' - a_n' b_n'| \\
&= |a_n| \cdot |b_n - b_n'| + |b_n'| \cdot |a_n - a_n'|
\end{aligned}
$$

$\square$

**Proof** Using Lemma 4.22, determine

$$
\begin{aligned}
p \quad & \text{such that } \forall n(|a_n|, |b_n| < p)\,, \\
n_1 \qquad & \forall m(|b_{n_1} - b_{n_1+m}| < (2^{-k-2})/p)\,, \\
n_2 \qquad & \forall m(|a_{n_2} - a_{n_2+m}| < (2^{-k-2})/p)\,.
\end{aligned}
$$

Set $n = \max(n_1, n_2)$. Now

$$
\begin{aligned}
|a_n| \cdot |b_n - b_n'| + |b_n'| \cdot |a_n - a_n'| &< p(|b_n - b_n'| + |a_n - a_n'|)\,, \\
&< p \left( \frac{2^{-k-2}}{p} + \frac{2^{-k-2}}{p} \right) = 2^{-k-1}\,.
\end{aligned}
$$

Thus, $(a_n b_n) \sim (a_n' b_n')$.                                                             $\square$

**Exercise 4.23** Prove Lemma 4.22.

**Exercise 4.24** Prove (1) and (3) of Lemma 4.21.

The definitions of the operations for the real numbers are now routine:

**Definition 4.25**

$$
\begin{aligned}
(a_n)/_\sim + (b_n)/_\sim &:= (a_n) + (b_n)/_\sim \\
(a_n)/_\sim \cdot (b_n)/_\sim &:= (a_n b_n)/_\sim \\
-(a_n)/_\sim &:= (-a_n)/_\sim
\end{aligned}
$$

The preceding lemma has shown that the operations on the real numbers are independent of the choices of the representatives. Moreover, we have

**Corollary 4.26**

(1)  $a = a' \wedge b = b' \rightarrow a + b = a' + b'$
(2)  $a = a' \wedge b = b' \rightarrow ab = a'b'$
(3)  $a = a' \rightarrow -a = -a'$

Now that we have defined the opposite of a real number, the difference of two real numbers is given by:

$$
a - b := a + (-b)
$$

Let us now give a list of properties that follow almost directly from the definitions:

(1) $a + b = b + a$
(2) $(a + b) + c = a + (b + c)$
(3) $ab = ba$
(4) $a(bc) = (ab)c$
(5) $a(b + c) = ab + ac$
(6) $a + 0 = a$
(7) $a \cdot 1 = a$
(8) $a + (-a) = 0$
(9) $-(-a) = a$
(10) $a + b > 0 \rightarrow a > 0 \vee b > 0$
(11) $ab > 0 \rightarrow (a > 0 \wedge b > 0) \vee (a < 0 \wedge b < 0)$

An important operation is still missing: namely, the inverse. In classical mathematics every real number $\neq 0$ has an inverse, i.e., $a \neq 0 \rightarrow \exists x(xa = 1)$; intuitionistically this is somewhat more difficult to establish.

To treat the inverse we introduce a positive analog to inequality. Inequality is a negative concept; it provides too little information. Roughly speaking, to calculate the inverse we need to find a Cauchy sequence that behaves like the sequence of the inverses of the terms of the given Cauchy sequence. Such a Cauchy sequence must remain bounded (otherwise it certainly cannot "converge"). But then "in the long run" the given Cauchy sequence must stay away from 0. A Cauchy sequence that is only unequal to 0 can still hang very close to 0 in an uncontrollable way.

**Definition 4.27**  $a \# b := a < b \vee b < a$

The relation $\#$ is called the *apartness relation*. Apartness comes down to having a positive distance, that is, a distance greater than a positive rational number.

Before we prove the properties of apartness, we first derive a few auxiliary theorems.

**Lemma 4.28**

*(1) If $r$ is a positive rational number, then $a < a + r$.*
*(2) $a > b \rightarrow \exists k(a > b + 2^{-k})$*
*(3) $a \leq b \leftrightarrow \forall k(a < b + 2^{-k})$*
*(4) $\forall k(-2^{-k} < a < 2^{-k}) \rightarrow a = 0$*

*Proof*

(1) Do it yourself.
(2) We know $a > b \leftrightarrow \exists nk \forall m(a_{n+m} - b_{n+m}) > 2^{-k}$. Take such $n$ and $k$, then $a_{n+m} - (b_{n+m} + 2^{-k-1}) > 2^{-k-1}$ for all $m$. By definition, it now follows that $a > b + 2^{-k-1}$, and therefore $\exists k(a > b + 2^{-k})$.
(3) $\rightarrow$ $b < b + 2^{-k}$, whence Theorem 4.14 yields $b < a$ or $a < b + 2^{-k}$. The first is excluded, so $a < b + 2^{-k}$ is true, for each $k$.

$\leftarrow$ Assume $b < a$. Now use the definition of $<$ and apply (2) to arrive at a contradiction.

(4) $a < 2^{-k} \leftrightarrow \exists p \exists n \forall m (2^{-k} - a_{n+m} > 2^{-p})$. Without loss of generality we can assume $p > k$. Then $\forall m (a_{n+m} < 2^{-k} - 2^{-p} < 2^{-k})$. Similarly, for a suitable $n$ we get $\forall m (a_{n+m} > -2^{-k})$. So for the maximum of the values found above for $n$ we get $\forall m (|a_{n+m}| < 2^{-k})$. This shows $(a_n) \sim \langle 0 \rangle$.

$\square$

## Theorem 4.29

*(1)* $a \leq b \wedge b \leq c \to a \leq c$
*(2)* $a < b \to a \leq b$
*(3)* $a \leq b \wedge b \leq a \to a = b$
*(4)* $a < b \wedge b \leq c \to a < c$

## *Proof*

(1) Do it yourself.
(2) This has already been proved in Theorem 4.12.
(3) $a \leq b \leftrightarrow \forall k(a < b + 2^{-k})$, and $b \leq a \leftrightarrow \forall k(b < a + 2^{-k})$. So $\forall k(-2^{-k} < b - a < 2^{-k})$. Apply Lemma 4.28: $b - a = 0$.
(4) From $a < b$ follows $c < b$ or $a < c$. The first possibility is excluded by the second assumption $b \leq c$, so $a < c$.

$\square$

**Exercise 4.30**  Show: $a > b \leftrightarrow \exists r \in \mathbb{Q} \, (a > r > b)$.

We now return to the apartness relation.

## Theorem 4.31

*(1)* $a \, \# \, b \to b \, \# \, a$
*(2)* $a \, \# \, b \to a \, \# \, c \vee c \, \# \, b$
*(3)* $\neg a \, \# \, b \leftrightarrow a = b$

## *Proof*

(1) Trivial.
(2) Apply Theorem 4.14.
(3) $\neg a \, \# \, b \leftrightarrow \neg(a < b \vee b < a)$
$\qquad\quad \leftrightarrow \neg a < b \wedge \neg b < a$
$\qquad\quad \leftrightarrow b \leq a \wedge a \leq b$
$\qquad\quad \leftrightarrow a = b$

$\square$

This isolates the most important properties of apartness. Property (2) is an especially useful tool in mathematics. If we think of # as "is at some distance from", then (2) says that if two points are at some distance from each other, and you pick an arbitrary point, then you know that it is at some distance from at least one of the other two points. In our definition, "apart" does not mean that the distance need be large, but just that it exceeds a certain $2^{-k}$; in experimental jargon "two numbers (or points) are *palpably different*" (a literal translation of Brouwer's Dutch term). Unlike $\neq$, # gives us positive information.

The apartness relation also relates to equality:

**Theorem 4.32** $\neg\neg a = b \rightarrow a = b$

**Proof** $\neg\neg a = b \leftrightarrow \neg\neg\neg a \mathbin{\#} b \leftrightarrow \neg a \mathbin{\#} b \leftrightarrow a = b$. $\qquad\qquad\qquad\square$

An equality relation that satisfies this theorem is called *stable*. Note that stability need not entail decidability, indeed here we have a weak counterexample, as equality of real numbers is not decidable.

**Exercise 4.33** $a \mathbin{\#} 0 \leftrightarrow \exists k n \forall m (|a_{n+m}| > 2^{-k})$

The apartness relation is indispensable when calculating inverses (that is, when dividing). In classical mathematics you can divide by a number as soon as you have the negative knowledge that it is not equal to 0, but in intuitionistic mathematics that is not enough. To be able to really calculate (some might say "approximate") the inverse via a Cauchy sequence, you must be able to estimate the distance from 0. Roughly speaking, if a sequence converges to 0, the inverses converge to infinity, and the terms of a Cauchy sequence must remain finite.

**Theorem 4.34** *a has an inverse* $\leftrightarrow a \mathbin{\#} 0$.

**Proof** $\rightarrow$. Let $ab = 1$ for a certain $b$, then $\forall k \exists n \forall m (|a_{n+m}b_{n+m} - 1| < 2^{-k})$. In particular, there is an $n_0$ such that for all $m$

$$|a_{n_0+m}b_{n_0+m} - 1| < \frac{1}{2}, \text{ i.e., } -\frac{1}{2} < a_{n_0+m}b_{n_0+m} - 1 < \frac{1}{2},$$

so $a_{n_0+m}b_{n_0+m} > 1/2$. The sequence $(b_n)$ is bounded, i.e., $\forall m (|b_m| < 2^p)$ for a certain $p$.

$$|a_{n_0+m}| \cdot |b_{n_0+m}| > \frac{1}{2}$$
$$|a_{n_0+m}| \cdot 2^p > \frac{1}{2}$$
$$|a_{n_0+m}| > 2^{-p-1}$$

By Exercise 4.33 this means $a \mathbin{\#} 0$.

$\leftarrow$. The obvious thing to do is to invert all terms of the given Cauchy sequence. Unfortunately, this is not always possible, for a number of terms 0 can occur. However, as $a \,\#\, 0$, there can only be finitely many such 0 terms, so things are not that bad. Create a new sequence $(c_n)$ that coincides with $(a_n)$, except where $a_n = 0$, where we now take $c_n$ to be 1. Furthermore, we can safely assume that $a$ is positive, and also that all $c_i$ are positive (why?). Claim:

1. $(c_n) \sim (a_n)$
2. $(c_n^{-1})$ is a Cauchy sequence.
3. $(a_n)(c_n) = 1$

The justification is as follows.

1. Since from a certain index on $a_n$ coincides with $c_n$, the equivalence is evident.
2. By (a) and the fact that $a \,\#\, 0$ it follows that $c \,\#\, 0$. Thus, for large $m$, $c_m$ is bounded away from 0, say $|c_m| > 2^{-q}$ for $m > n_1$. Furthermore, since $(c_n)$ is a Cauchy sequence, $\forall p \exists n_2 \forall m (|c_{n_2} - c_{n_2+m}| < 2^{-p})$. Take $n = \max(n_1, n_2)$. For $m > n$ we have

$$|c_n^{-1} - c_m^{-1}| = |\frac{c_m - c_n}{c_n c_m}| = \frac{|c_m - c_n|}{|c_n| \cdot |c_m|},$$
$$< \frac{|c_m - c_n|}{2^q 2^q} < \frac{2^{-p}}{2^{2q}} = 2^{-p-2q}.$$

   Thus, $(c_n^{-1})$ is a Cauchy sequence.
3. Prove $(a_n)(c_n^{-1}) \sim 1$ yourself.

$\square$

Finally, we note that there is *at most* one inverse: $ab = ac = 1 \rightarrow b = c$. One sees that $1 = ac \rightarrow b = b(ac) = (ab)c = c$.

The links between apartness and the operations are also important. Here is multiplication:

**Theorem 4.35**  $ab \,\#\, 0 \leftrightarrow a \,\#\, 0 \wedge b \,\#\, 0$.

**Proof** $\rightarrow$. If $ab \,\#\, 0 \rightarrow ab$ has an inverse, say $c$, then $ab \cdot c = 1$, and $a(bc) = 1$; it follows that $a$ has an inverse, and is thus removed from 0. Ditto $b \,\#\, 0$.
$\leftarrow$. Call $c$ the *inverse* of $a$, and $d$ the inverse of $b$. Then $ac = 1, bd = 1$. And also $ac \cdot bd = 1$, and then $ab \cdot cd = 1$. We see that $ab$ has an inverse, so $ab \,\#\, 0$.  $\square$

From this theorem follows immediately $ab = 0 \leftrightarrow \neg ab \,\#\, 0 \leftrightarrow \neg(a \,\#\, 0 \wedge b \,\#\, 0)$, but no more. That is, we cannot assert $ab = 0 \leftrightarrow a = 0 \vee b = 0$.

We can introduce as many operations as we think are convenient. Here are a few:

**Definition 4.36** (*absolute value, maximum and minimum*)

(1)  $|(a_n)| := (|a_n|)$
(2)  $\max((a_n), (b_n)) := (\max(a_n, b_n))$
(3)  $\min((a_n), (b_n)) := (\min(a_n, b_n))$

**Lemma 4.37**  *The sequences* $|(a_n)|, \max((a_n), (b_n)), \min((a_n), (b_n))$ *are Cauchy-sequences.*

**Theorem 4.38**

*(1)  $|a + b| \leq |a| + |b|$ (the "triangle inequality")*
*(2)  $|ab| = |a| \cdot |b|$*
*(3)  $|a| = \max(a, -a)$*
*(4)  $a \geq 0 \leftrightarrow a = |a|$*

**Exercise 4.39**  Prove Theorem 4.38.

What is the relationship between a Cauchy sequence $a$ and its elements? It is obvious that the distance between the $\langle a_n \rangle$ (Definition 4.19) and $a$ can become arbitrarily small. The proof is not difficult, but we have to do proper accounting, that is, be careful whether we are talking about real numbers or rational numbers.

**Exercise 4.40**  Show that for any Cauchy sequence $a$, $\forall k \exists n \forall m > n(|a - \langle a_m \rangle| < 2^{-k})$.

Not all properties of the operations in classical mathematics can be transferred to intuitionistic mathematics. Here are two weak counterexamples.

1.  $ab = 0 \to a = 0 \vee b = 0$ does not hold.
    Consider the Cauchy sequences $(a_n)$ and $(b_n)$ with

$$a_n = \begin{cases} 2^{-n} \text{ if } \forall p < n \neg N_{20}(p) \\ 2^{-p} \text{ if } p < n \wedge N_{20}(p) \wedge (\forall m < p) \neg N_{20}(m) \end{cases}$$

$$b_n = \begin{cases} 1 \text{ if } \forall p < n \neg N_{20}(p) \\ 0 \text{ if } p < n \wedge N_{20}(p) \wedge (\forall m < p) \neg N_{20}(m) \end{cases}$$

    The predicate $N_{20}(n)$ was defined in Sect. 4.5: "the $n$-th decimal after the point of $\pi$ is preceded by a sequence of 20 nines". So in words the definition above says: write down the decimals of $\pi$ and at the same time the terms of the sequences $(a_n)$ and $(b_n)$. As long as no sequence of 20 nines has occurred in $\pi$, $a_n$ is equal to $2^{-n}$ and $b_n$ to 1; if at $p$ such a sequence occurs, then $(a_n)$ becomes constant $2^{-p}$ and $(b_n)$ becomes constant 0. Obviously we produce two Cauchy sequences $a$ and $b$ for which $(a_n b_n) \sim 0$. However, for the moment we cannot tell whether $a = 0$ or $b = 0$.
2.  $\max(a, b) = a \vee \max(a, b) = b$ not valid.
    Consider $\max(a, 0)$, where

$$a_n = \begin{cases} (-2)^{-n} \text{ if } \forall p < n \neg N_{20}(p) \\ (-2)^{-p} \text{ if } N_{20}(p) \wedge (\forall m < p) \neg N_{20}(m). \end{cases}$$

**Exercise 4.41**  Prove, for real numbers $a, b, c, d$:

(1)  $\min(a, b) \le a, b \le \max(a, b)$
(2)  $\min(a, b) = \max(a, b) \to a = b$
(3)  $\max(a, b) + \min(a, b) = a + b$
(4)  $c \ge \max(a, b) \leftrightarrow c \ge a \wedge c \ge b$
(5)  $|a| - |b| \le |a - b|$ (the "reverse triangle inequality")
(6)  $|a| = |-a|$
(7)  $a + b \mathbin{\#} 0 \to a \mathbin{\#} 0 \vee b \mathbin{\#} 0$
(8)  $ab \mathbin{\#} 1 \to a \mathbin{\#} 1 \vee b \mathbin{\#} 1$
(9)  $a + b \mathbin{\#} c + d \to a \mathbin{\#} c \vee b \mathbin{\#} d$
(10)  $a + b < c + d \to a < c \vee b < d$
(11)  $a \mathbin{\#} b \to a + c \mathbin{\#} b + c$

## 4.7  Limits

The real numbers are given as Cauchy sequences of rational numbers. This provided
an extension of the rational numbers with "the missing limits". We could now try to
go one step further by adding to the real numbers all the missing limits again. More
precisely, we can apply the Cauchy procedure one more time, this time to the real
numbers. That way we at least get an extension, but is it a real one? The answer
is no: all limits of the new Cauchy sequences are already there. The topological
terminology is "the Cauchy real numbers are *complete*". The proof is a matter of
neat accounting.

Here is a definition for convenience.

**Definition 4.42**  (i) A sequence of real numbers $(a_n)$ is called *convergent* if there is a
real number $a$ such that $\forall k \exists n \forall m > n(|a - a_m| < 2^{-k})$. We say that $(a_n)$ *converges to*
$a$ and call the number $a$ the *limit* of the sequence $(a_n)$. Symbolically: $\lim_{n \to \infty} a_n = a$.

Convergence is highly sensitive to the logical context. Given that in constructive
mathematics we allow fewer logical principles than in classical mathematics, there
is a chance that certain convergence theorems no longer go through. We give an
example: *a bounded monotone ascending (or non-descending) sequence converges*.
The classical proof is simplicity itself: the set $\{a_n \mid n \in \mathbb{N}\}$ is bounded and thus has
a supremum. It's a simple routine to show that this supremum is the limit of the
sequence. Now consider the following sequence $(a_n)$:

$$
a_n = \begin{cases} -2^{-n} & \text{if } \forall p < n \neg N_{20}(p) \\ 1 - 2^{-n} & \text{if } n \ge p \wedge N_{20}(p) \wedge (\forall m < p) \neg N_{20}(m) \end{cases}
$$

The sequence approaches 0 as long as there has not yet been a sequence of 20
nines in $\pi$; thereafter the sequence approaches 1. Suppose $(a_n)$ has a limit $a$, then
$a \mathbin{\#} 0 \vee a \mathbin{\#} 1$. From this it follows that a sequence of 20 nines either occurs in $\pi$

or it does not. But that has not been decided yet, and so convergence has not been demonstrated.

This shows that in constructive mathematics we have a weak counterexample to the *Bolzano-Weierstrass Theorem*, which says that in a closed interval an infinite set of points has a limit point.

**Theorem 4.43**

$$(a_n) \text{ is a Cauchy sequence in } \mathbb{R} \leftrightarrow (a_n) \text{ converges},$$

$$\leftrightarrow \lim a_n \text{ exists in } \mathbb{R}.$$

**Proof**

$\leftarrow$ is simple—do it yourself.

$\rightarrow$: The Cauchy condition for the sequence of real numbers $(a_n)$ is

$$\forall k \exists n \forall m > n (|a_n - a_m| < 2^{-k})$$

We now use the relationship between $n$ and $k$ to "compress" the sequence. Determine

$$a_{n_0} \text{ such that } \forall m > n_0 (|a_{n_0} - a_m| < 2^{-0})$$

$$\vdots$$

$$a_{n_k} \qquad \forall m > n_k (|a_{n_k} - a_m| < 2^{-k})$$

$$\vdots$$

Set $b_k := a_{n_k}$. Obviously $(b_n)$ is itself a Cauchy sequence. We are going through the elements of the Cauchy sequence $(a_n)$ in order, but with jumps, each time to an element of $(a_n)$ from which on $(a_n)$ fulfills the Cauchy condition for the next smaller power of 2.[7] So the sequence $(b_n)$ is a *subsequence* of the given Cauchy sequence, and it satisfies $|b_n - b_m| < 2^{-n}$ for $m > n$. We now concentrate on $(b_n)$. Every $b_n$ is itself a Cauchy sequence of rational numbers:

$$b_0 = b_{00}, b_{01}, b_{02}, b_{03}, \ldots$$
$$b_1 = b_{10}, b_{11}, b_{12}, b_{13}, \ldots$$
$$b_2 = b_{20}, b_{21}, b_{22}, b_{23}, \ldots$$
$$\vdots$$
$$b_n = b_{n0}, b_{n1}, b_{n2}, b_{n3}, \ldots$$
$$\vdots$$

We are now going to determine a tailpiece in every $b_n$ in which the differences are smaller than $2^{-n}$.

---

[7] Depending on the method we have to determine that element, the latter need not be the *first* element from which on that condition is fulfilled; but that is immaterial.

Choose an $n$ and consider the tailpiece $b_{n,p_n}, b_{n,p_n+1}, b_{n,p_n+2}, \ldots$ with the property $\forall m > p_n (|b_{n,p_n} - b_{n,m}| < 2^{-n})$. We know that $|b_k - b_s| < 2^{-k}$ for $s > k$; furthermore, we can always choose the $p_n$ such that $|b_n - \langle b_{n,p_n}\rangle| < 2^{-k}$ (recall from Definition 4.19 that $\langle b_{n,p_n}\rangle$ is the embedding of the rational number $b_{n,p_n}$ into the real numbers; and see Exercise 4.5).

$$\begin{aligned}
b_0 &= b_{00}, \ldots, b_{0p_0}, \ldots\\
b_1 &= b_{10}, b_{11}, \ldots\ldots, b_{1p_1}, \ldots\\
b_2 &= b_{20}, b_{21}, \ldots\ldots\ldots, b_{2p_2}, \ldots\\
&\ \ \vdots\\
b_n &= b_{n0}, b_{n1}, b_{n2}, \ldots\ldots\ldots\ldots, b_{np_n}, \ldots\\
&\ \ \vdots
\end{aligned}$$

We show that this diagonal-like sequence $(b_{np_n})$ is a Cauchy sequence. Consider $|\langle b_{np_n}\rangle - \langle b_{s,p_s}\rangle|$:

$$\begin{aligned}
|\langle b_{n,p_n}\rangle - \langle b_{s,p_s}\rangle| &\leq |\langle b_{n,p_n}\rangle - b_n| + |b_n - b_s| + |b_s - \langle b_{s,p_s}\rangle|,\\
&< 2^{-n} + 2^{-n} + 2^{-s} < 2^{-n+2}.
\end{aligned}$$

From this immediately follows $|b_{np_n} - b_{s,p_s}| < 2^{-n}$, so the sequence $(b_{np_n})$ is a Cauchy sequence, say $b$. It is now not difficult to show that $(b_n)$ converges to $b$ (do it yourself). $\qquad\square$

We now turn to the most important properties of limits. For the most part they are not difficult to prove. We assume that when we use $\lim_{n\to\infty} a_n$, that limit also exists. In exceptional cases we indicate what may or may not be assumed. When there is no danger of misunderstanding we usually write $\lim a_n$.

**Theorem 4.44**

*(1)* $\lim(a_n + b_n) = \lim a_n + \lim b_n$
*(2)* $\lim(a_n \cdot b_n) = \lim a_n \cdot \lim b_n$
*(3)* $\lim \max(a_n, b_n) = \max(\lim a_n, \lim b_n)$
*(4)* *If $\forall n(a_n \mathbin{\#} 0)$ and $\lim a_n^{-1}$ exists, then* $\lim a_n^{-1} = (\lim a_n)^{-1}$
*(5)* $\forall n(a_n \leq b_n) \to \lim a_n \leq \lim b_n$

**Exercise 4.45** Prove Theorem 4.44.

Limits are often important tools for defining special numbers. Convergent series are even more useful. There is an unlimited amount to say and prove about both topics. Because here we introduce the main topics, and in doing so want to show that the usual tools of analysis are perfectly usable (albeit with some precaution), we limit ourselves to a number of concepts and properties.

We consider sequences of real numbers.

**Definition 4.46** $s_n = \sum_{i=1}^{n} a_n$ is called the *n-th partial sum* of the sequence $(a_n)$. We call $\lim s_n$ the *sum of the series*, which we denote as $\sum_{n=1}^{\infty} a_n$, or $\sum_{1}^{\infty} a_n$. If the limit exists, we call the series *convergent*. The series is called *absolutely convergent* if the series of the absolute values converges.[8]

**Exercise 4.47** Prove that if the series $\sum a_n$ converges then $\lim a_n = 0$.

In classical analysis there is a plethora of techniques to determine the convergence of series. We show, without aiming for completeness, that certain tricks also work in constructive analysis.

**Theorem 4.48** (Comparison Test) *If* $\forall n (b_n \geq 0)$, $\sum b_n$ *converges and* $\forall n |a_n| \leq b_n$, *then* $\sum a_n$ *converges and also converges absolutely.*

**Proof** The series $\sum b_n$ converges, and is thus a Cauchy sequence, i.e. $\forall k \exists n \forall m > n \sum_{i=n}^{m} b_i < 2^{-k}$. Now we have:

$$\left| \sum_{i=n}^{m} a_i \right| \leq \sum_{i=n}^{m} |a_i| \leq \sum_{i=n}^{m} b_i < 2^{-k}$$

We see that $\sum a_n$ and $\sum |a_n|$ are Cauchy sequences, so they converge. $\qquad \square$

There is also a positive definition of divergence, not just non-convergence.

**Definition 4.49** The sequence $(a_n)$ *diverges* if $\exists p \forall k \exists m > k \exists n > k (|a_n - a_m| > 2^{-p})$. A series *diverges* if the sequence of the partial sums diverges.

In words, no matter how far you go in the line, there are always two terms that are further than a predetermined distance from each other.
Examples: (1) the series $\sum 1/\sqrt{n}$ diverges. Note that

$$\frac{1}{\sqrt{n}} + \frac{1}{\sqrt{n+1}} + \cdots + \frac{1}{\sqrt{n^2}} > (n^2 - n + 1)\frac{1}{\sqrt{n^2}} > \frac{n^2 - n + 1}{n} > n - 1 + \frac{1}{n} > n - 1.$$

(2) The series $\sum (-1)^{n+1}/n$ converges. Take together consecutive pairs in the series $1 - 1/2 + 1/3 - 1/4 + \cdots$ and estimate a sufficiently long tail $a_n + \cdots + a_{2n}$.

$$\frac{1}{2n-1} - \frac{1}{2n} + \cdots \frac{1}{4n-1} - \frac{1}{4n} = \frac{1}{4n^2 - 2n} + \cdots + \frac{1}{4(2n)^2 - 2 \cdot 2n}$$

$$< \frac{n+1}{4n^2 - 2n}$$

$$< \frac{n}{4n^2 - n^2} = \frac{1}{3n}.$$

---

[8] The difference between "sequence" and "series" seems a bit esoteric. One can say that the series associated with the sequence $(a_n)$ is the sequence of partial sums.

Show that the series does not converge absolutely. Such a series is called *conditionally convergent*.

**Exercise 4.50**  (i) Prove that the series $\sum 1/n$ diverges.
(ii) Prove that a series diverges if there are infinitely many terms greater than a given positive number.

**Theorem 4.51**  (Ratio Test) *Let $c > 0$, $n \in \mathbb{N}$, then $\sum a_m$ converges if $c < 1$ and*
$\forall m > n\, |a_{m+1}| \le c |a_m|$.
*If $c > 1$ and $\forall m > n\, |a_{m+1}| > c |a_m|$, then $\sum a_n$ diverges.*

**Proof** Take $c < 1$, then $\forall m > n\, |a_m| \le c^{m-n} |a_n|$. Now apply the Comparison test. Do the part about divergence yourself.                                                                $\square$

We now consider series and sequences of continuous functions on a closed interval. For the theory it doesn't matter much which interval we take, say $[0, 1]$.

**Definition 4.52**  The sequence $(f_n)$ *converges on* $[0, 1]$ *to the continuous $f$ if*

$$\forall k\ \exists n\ \forall m > n\ \forall x \in [0, 1]\ (|f_n(x) - f(x)| < 2^{-k})\,.$$

The sequence $(f_n)$ is *Cauchy on* $[0, 1]$ if

$$\forall k\ \exists n\ \forall m > n\ \forall x \in [0, 1]\ (|f_n(x) - f_m(x)| < 2^{-k})\,.$$

The theory of sequences of functions behaves largely analogously to that of sequences of numbers. We therefore omit the proofs if they can be easily imitated.

**Theorem 4.53**  $(f_n)$ *converges* $\leftrightarrow$ $(f_n)$ *is a Cauchy sequence.*

**Theorem 4.54**  *The Comparison Test applies to function sequences:*
*if $\sum g_n$ converges, where the functions $g_n$ on $[0, 1]$ are not negative, and*

$$\forall x\, (|f_n(x)| \le |g_n(x)|)\,,$$

*then $\sum f_n$ converges.*

The next statement deals with a slightly different domain.

**Theorem 4.55**  *The Ratio Test applies to function series: if on every closed subinterval of $[0, 1]$ $f_n$ meets the condition*

$$\exists c \in (0, 1)\ \exists n\ \forall m > n\ \forall x\ (|f_{m+1}| \le c |f_m(x)|)\,,$$

*then $\sum f_n$ absolutely converges on $[0, 1]$.*

**Definition 4.56**  A series of the form $\sum a_n (x - u)^n$ is called a *power series*.

A simple application of the Ratio Test proves

**Theorem 4.57**  *If for $(a_n)$ we have*

$$\exists r > 0 \, \exists n \, \forall m > n \, (|a_{m+1}| \leq (1/r)|a_m|) \,,$$

*then the series $\sum a_n (x - u)^n$ converges absolutely on the interval $(u - r, u + r)$.*

***Proof***  Take a closed sub-interval $[s, t]$ of $(u - r, u + r)$, then for all $x$ in that interval we get $|x - u| \leq r_0$ for an $r_0$ with $0 < r_0 < r$.

$$|a_{m+1}| \leq \frac{1}{r}|a_m| \rightarrow |a_{m+1}(x - u)^{m+1}| \leq \frac{1}{r}|a_m(x - u)^{m+1}| \leq \frac{r_0}{r}|a_m(x - u)^m| \,.$$

According to the Ratio Test, the series converges absolutely on $(u - r, u + r)$.   $\square$


## 4.8  Supremum and Infimum

So far we have avoided using sets. Since they play a prominent role, we will now take a closer look at them. We consider subsets of known sets or spaces, say $\mathbb{N}$ or $\mathbb{R}$.

**Definition 4.58**  The set $X$ is *empty* $:= \forall x \neg (x \in X)$.

If one accepts the *Ex Falso* principle,[9] one can prove, as in classical mathematics, that there is exactly one empty set; we denote it with $\emptyset$.

In constructive circles, various sets with other definitions are regarded with some justified suspicion. The following examples show why. $R$ is the Riemann Hypothesis.

(1)  $X_1 = \{n \in \mathbb{N} \mid (n = 0 \wedge R) \vee (n = 1 \wedge \neg R)\}$.
(2)  $X_2 = \{n \in \mathbb{N} \mid R\}$.
(3)  $X_3 = \{n \in \mathbb{N} \mid n \leq 9 \text{ and } n \text{ occurs infinitely often in the decimal expansion of } \pi\}$.

Classically, the set $X_1$ is either $\{0\}$ or $\{1\}$, depending on whether the Riemann Hypothesis is correct or not. We just don't know yet which of the two it is. The second set either contains exactly all natural numbers, or nothing, again depending on the correctness of $R$. The set $X_3$ is at present totally unknown, it can be any non-empty subset of $\{0, 1, \ldots, 9\}$. Constructively, sets of this kind are so annoying because we can't tell where they are located, what's in them, etc. Evidently, some sets have a high degree of undecidability.

The above examples show, among other things, that a set's being "non-empty" does not tell us that much. Consider $X_1$ again. If $X_1$ were $\emptyset$, then neither 0 nor 1 would belong to it, i.e., $R$ and its negation would both be false. And that is out of the question. So $X_1$ is non-empty, but we cannot say that it contains an element until we have decided $R$. So we have a weak counterexample to $\neg (X_1 = \emptyset) \rightarrow \exists x (x \in X_1)$;

---

[9] See Theorem 2.1 and its proof for the classical reading, and Sect. 2.2 for some intuitionistic considerations.

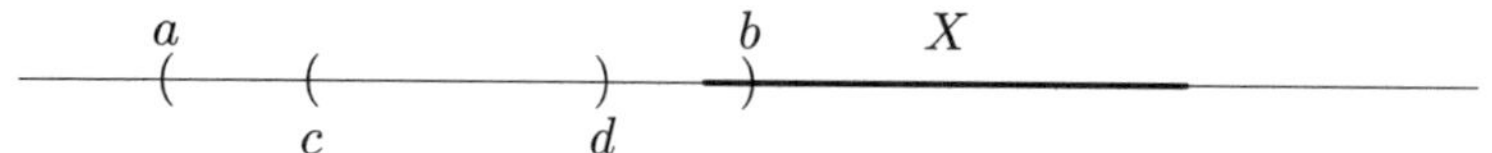

**Fig. 4.6** Locating $X$ by using not one, but two intervals

from the definition of $\emptyset$ it is clear that in fact we already saw this one at the end of Sect. 2.2.

Therefore, we introduce a more useful notion:

**Definition 4.59** The set $X$ is *inhabited* $:= \exists x (x \in X)$.

Clearly, "uninhabited" is the same as empty, but "inhabited" tells us more than "non-empty".

Since untidy sets do occur in mathematical practice, constructivists must always beware of strange effects. Brouwer recognized early on that there is a striking difference between sets for which one has a method to construct all their elements, and sets that are defined merely by some property that their elements have. For the latter he introduced a separate name: *species*.[10]

For us, sets of real numbers are the most interesting, especially because analysis uses them all the time.

While with the natural numbers (and likewise with the integers and rational numbers) we have a good idea of what "neat" sets are—namely the decidable ones, the real numbers present us with a problem. Decidability plays no role. We have already seen that in general it is not possible to tell whether two real numbers are equal or not. What then can one expect from sets? Think of two real numbers $a$ and $b$ of which the equality is unknown, what do we know about $\{a, b\}$? Does it have two elements, or one? And is the set $\{7, \pi\}$ decidable, i.e., can it be determined for every real number $a$ whether it is included? As long as there will be weak counterexamples, this is not the case.

We clearly need some alternative concept of "niceness" of subsets of $\mathbb{R}$. There is indeed such a concept; Brouwer introduced it in 1918, and it has been a standard part of constructive mathematics ever since. Its definition can be further motivated as follows.

While it would be asking too much to insist on decidability of $a \in X$ for subsets $X$ of $\mathbb{R}$, it would help if we could determine at least something about where $X$ is lying among the reals, for example whether a test interval overlaps $X$. Unfortunately, that is still asking too much. Consider $\{0\}$ and $(a, b)$, where $a < b$ and for $a$ it is unknown where it lies with respect to 0. Then it is unknown whether $\{0\} \cap (a, b)$ is empty.

The solution is to run the test with two nested intervals $(a, b)$ and $(c, d)$, where $a < c < d < b$. In Fig. 4.6 we have drawn a neat subset $X$. We see that the inner interval does not touch $X$, but the outer does. Now when for all such interval pairs

---

[10] On species, see also Sect. 6.1 below.

the outermost touches the set or the innermost does not, then the set $X$ can be located very neatly; the freedom of $X$ to jump is then sufficiently limited.

**Definition 4.60**  $X \subseteq \mathbb{R}$ is *located* $:=$ for every $a < c < d < b$, $(a, b) \cap X$ is inhabited or $(c, d) \cap X = \emptyset$.

**Theorem 4.61**

*(1)  $\mathbb{R}$ and $\emptyset$ are located.*
*(2)  Every singleton $\{t\}$ is located.*
*(3)  Each pair $\{t, s\}, t \neq s$ is located.*

***Proof***  (1) is immediate from the definition.

We see (2) as follows. Let $t \in \mathbb{R}$ and choose $a < c < d < b$. We now apply the law $x < y \rightarrow x < z \vee z < y$ to $a < c$: $a < t \vee t < c$. In the case $t < c$ we are done (i.e., $\{t\} \cap (c, d) = \emptyset$). Because $d < b$, now holds $t < b \vee d < t$. In the second case, we're done. In the first case, with the preceding $a < t \wedge t < b$, i.e., $t \in (a, b)$.

For (3) we use a similar line of reasoning. Let $t, s \in \mathbb{R}$. With an argument like the above we find
$$t \in (a, b) \vee t \notin (c, d)$$
$$s \in (a, b) \vee s \notin (c, d)$$

Now there are four possibilities:

$$t \in (a, b) \wedge s \in (a, b)$$
$$t \in (a, b) \wedge s \notin (c, d)$$
$$t \notin (c, d) \wedge s \in (a, b)$$
$$t \notin (c, d) \wedge s \notin (c, d)$$

Of these, the first three yield: $\{t, s\} \cap \{a, b\}$ is inhabited, and the last one yields $\{t, s\} \cap \{a, b\} = \emptyset$. $\qquad\qquad\square$

(3) also follows from (2) as a special case of

**Theorem 4.62**  *If $X$ and $Y$ are located, then so is $X \cup Y$.*

***Proof***  Mimic the above proof. $\qquad\qquad\square$

By applying induction to the cases (2) and (3)—or by repeatedly applying (2) and the above theorem—one gets all finite cases. We now must make a refinement. In case (3), it is not at all certain that we are dealing with a finite set; as long as we do not know whether $s$ and $t$ are different, we cannot bring $\{t, s\}$ into an unambiguous correspondence with a natural number. We only know that there is a function $f$ on $\{0.1\}$ with $f(0) = t$, $f(1) = s$. The set $\{t, s\}$ is, so to speak, "indexed" by $\{0, 1\}$.

**Definition 4.63**  A set $X$ is called *finite* if there is an $n$ and a bijection of $\{0.1, \ldots, n - 1\}$ onto $X$. A set $X$ is called *index-finite* if there is an $n$ and an image of $\{0, 1, \ldots, n - 1\}$ on $X$.

Note that finite entails index-finite, but the converse is not true—give a weak counterexample.

So we have:

**Corollary 4.64** *An index-finite set is located.*

Unfortunately, for located sets $X$ and $Y$, we don't get that $X \cap Y$ is located. Consider only two numbers $s$ and $t$ of which it is unknown whether they are equal and form $A = \{s\} \cap \{t\}$. Since we can't decide if $A$ is inhabited or empty, we can't run the locatedness test.

The term "locatedness" can also be used to cover a few important concepts from analysis: supremum and infimum.[11]

**Definition 4.65** $a$ is called the *supremum* of X ($a = \sup(X)$) if

(1)  $\forall x \in X \ (x \leq a)$
(2)  $\forall k \ \exists x \in X \ (a - x < 2^{-k})$

$b$ is called the *infimum* of X ($b = \inf(X)$) if

(1)  $\forall x \in X \ (b \leq x)$
(2)  $\forall k \ \exists x \in X \ (x - b < 2^{-k})$

There is a well-known theorem from traditional analysis that any subset of $\mathbb{R}$ that is bounded from above has a supremum. The statement is constructively incorrect. Here is a weak counterexample: $X = \{n \in \mathbb{N} \mid n = 0 \vee (n = 1 \wedge R) \vee (n = 2 \wedge \neg R)\}$, where again $R$ is the Riemann Hypothesis. $X$ is inhabited (so it is not an empty set) and it is bounded from above. Now suppose that $X$ had a supremum $a$, then $a > 1 \vee a < 2$ would hold. In the first case $R$ would be incorrect and in the second case $\neg R$ would be incorrect. Neither is known yet. So $\sup(X)$ does not exist yet. We could also have used the number $a$ we constructed in Chap. 1, based on an undecided question about the decimals in $\pi$: at present, we cannot construct a supremum of $\{0, a\}$.

Inhabited located sets have a sup and an inf. However, we can weaken the condition a bit. For a sup, so to speak, only the upper region of a set needs to be sufficiently clear.

**Theorem 4.66** *Let $X \subseteq \mathbb{R}$ be inhabited and bounded above. If for all $a < b$, either $(b, \infty) \cap X = \emptyset$ or $(a, \infty) \cap X$ is inhabited, then $X$ has a supremum.*

Here $(a, \infty)$ is the set of all numbers greater than $a$. Another way to formulate the same condition is: *b is an upper bound or a is less than an element of* $X$.

***Proof*** See Fig. 4.7. $X$ contains at least one element, say $a_0$. Furthermore, there is an upper bound, say $b_0$. We are now going to approach the supremum. Divide the interval $(a_0, b_0)$ into three equal parts $(a_0, t), (t, s), (s, b_0)$. According to the condition of the

---

[11] Also called *least upper bound* and *greatest lower bound*. The contractions *upper boundary* and *lower boundary* are less felicitous.

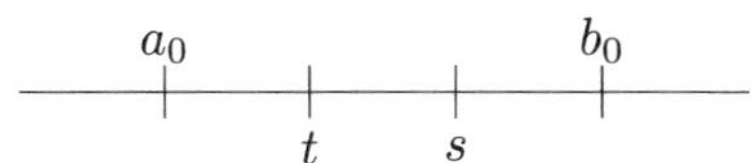

**Fig. 4.7**  Approaching the supremum of a set that contains $a_0$ and is bounded by $b_0$

theorem, $s$ is an upper bound or $t$ is less than an element of $X$, i.e., $\exists x \in X\ (t < x)$. In the first case we set $a_1 = a_0$, $b_1 = s$, and in the second case $a_1 = t$, $b_1 = b_0$. The length of the interval $(a_1, b_1)$ is $(2/3)u$. Now repeat the procedure with $(a_1, b_1)$, which yields $(a_2, b_2)$ with length $(2/3)^2 u$. Continuing like this we get a sequence of contracting intervals $(a_n, b_n)$ with length $(2/3)^n u$. Define $w = \lim_{n \to \infty} a_n$. (So also $w = \lim_{n \to \infty} b_n$.) Claim: $w$ is $\sup(X)$. To see this, take an $x \in X$, then $x \leq b_n$ for all $n$. So $x \leq \lim(b_n)$. Choose a $k$; we need to find an $x \in X$ that lies less than $2^{-k}$ below $w$. First determine an $a_n$ with $w - a_n < 2^{-k}$ (remember that $(a_n)$ is non-decreasing). Based on the definition of the $a_n$'s there is above $a_n$ an element $x \in X$, and $x \leq w$, so $w - x < 2^{-k}$ holds. $\qquad\square$

**Exercise 4.67**  Show that if $X$ has a supremum, the conditions of Theorem 4.66 are satisfied.

Wholly analogously to the above we have

**Theorem 4.68**  *If $X \subseteq \mathbb{R}$ is inhabited and bounded from below, then $X$ has an infimum if for every pair $a$, $b$ for which $a < b$ holds, we have: Either $(-\infty, a) \cap X = \emptyset$, or $(-\infty, b) \cap X$ is inhabited.*

It now follows directly from the two previous statements:

**Theorem 4.69**  *If $X$ is an inhabited, bounded, located set, then $X$ has an infimum and a supremum.*

Note: An adapted characterization also applies, of course, in subspaces such as $[0,1]$ or $[0, \infty)$.

**Exercise 4.70**  Prove $\max(a, b) = (\max(a_n, b_n))_n = \sup\{a, b\}$.

Our definition of "located" is based on comparison with intervals. We can also use the concept of "distance from a point to a set".

**Definition 4.71**  The *distance* from a point $a$ to a set $X$ is defined as $d(a, X) = \inf\{|a - x| \mid x \in X\}$.

**Theorem 4.72**  *If $X$ is inhabited then the following holds: $X$ is located $\leftrightarrow d(a, X)$ exists for all $a$.*

**Proof** $\rightarrow$: for $a$ consider the intervals $(a - u, a + u)$ and $(a - v, a + v)$, where $u > v > 0$. Because of the locatedness of $X$, $(a - u, a + u) \cap X$ is inhabited or $(a - v, a + v) \cap X = \emptyset$. So for the set $Y$ of distances from $a$ to $X$, $\{|a - x| \mid x \in X\}$, it follows that, for $u$ and $v$, $Y \cap (0, u)$ is inhabited or $(0, v) \cap Y = \emptyset$. That is, $Y$ satisfies the condition for the existence of an infimum, so $d(a, X)$ exists.

$\leftarrow$: Let $u_0 < v_0 < v_1 < u_1$ and choose $a = (v_0 + v_1)/2$. Set $r_1 := a - v_0 = v_1 - a$, $r_2 := a - u_0$, $r_3 := u_1 - a$. We have $r_1 < r_2$ and $r_1 < r_3$, so $r_1 < d(a, X) \vee d(a, X) < r_2$ and $r_1 < d(a, X) \vee d(a, X) < r_3$. If $r_1 < d(a, X)$, then $X \cap (v_0, v_1) = \emptyset$. There remains the case $d(a, X) < r_2$ and $d(a, X) < r_3$ to consider. From the definition of distance it now easily follows that $X \cap (u_0, u_1)$ is inhabited. This shows that $X$ is located. $\qquad\qquad\square$

There is another concept that plays an important role in constructive mathematics.

**Definition 4.73** A set $X$ is called *totally bounded* if for every $k$ there is a finite set of points $p_0, p_1, \ldots, p_{n-1}$ in $X$ with

$$\forall x \in X \; \exists i < n \; (|x - p_i| < 2^{-k}).$$

The concept also occurs in classical topology: such a set of approximating points is called an $\epsilon$-*net* (in our context a $2^{-k}$-net).

In constructive mathematics, "totally bounded and closed" (or "complete") takes the place of the classical "compact".

**Theorem 4.74** *$X$ is totally bounded $\rightarrow$ $X$ is located.*

**Proof** Consider the test for $s_1 < s_2 < s_3 < s_4$, then we find a middle interval that is a little larger: take $k$ such that $s_1 < s_2 - 2^{-k} < s_3 + 2^{-k} < s_4$. Then determine the approximating set $P = \{p_0, p_1, \ldots, p_{n-1}\}$ for $2^{-k}$. $P$ is located, so $P \cap (s_1, s_4)$ is inhabited or $P \cap (s_2 - 2^{-k}, s_3 + 2^{-k}) = \emptyset$. In the first case $X \cap (s_1, s_4)$ is inhabited; the second case we have to look further into. Set $w \in X \cap (s_2, s_3)$ for a certain $w$, then there is a $p_i$ with $|p_i - w| < 2^{-k}$, but then $p_i \in (s_2 - 2^{-k}, s_3 + 2^{-k})$. Contradiction. Ergo $X \cap (s_2, s_3) = \emptyset$. $\qquad\qquad\square$

Finally, a theorem about the real numbers themselves. In classical mathematics one can prove that $\mathbb{R}$ is connected, that is, that $\mathbb{R}$ cannot be the disjoint union of two open sets. That is a negative statement. The corresponding positive statement is classically equivalent: Two open sets that together span the continuum (i.e., $\mathbb{R}$) have a non-empty intersection (still negative wording). We will show that in constructive mathematics there is a wholly positive version.

**Theorem 4.75** (Positive connectedness) *If $[a, b] = A \cup B$ for two inhabited open sets $A$ and $B$, then there is a $c \in A \cap B$.*

**Proof** We recall that an open subset of $\mathbb{R}$ has by definition the property that every point of it is contained in an open interval within that set.

The proof consists of an approximation. Pick any point $x$ in $A$ and any point $y$ in $B$. Since $A$ is open, there is a $k$ such that $x + 2^{-k} \in A$. Now it follows from $x \,\#\, x + 2^{-k}$ that $y$ is apart from $x$ or from $x + 2^{-k}$. So there are two points, the one in $A$ and the other in $B$, that are apart from one another. Without loss of generality we can assume that such a point of $B$ is to the right of such a point of $A$. So choose $a_0 \in A$ and $b_0 \in B$ with $a_0 < b_0$.

Set $u := b_0 - a_0$ and $c_0 := 2^{-1}(a_0 + b_0)$. Then $c_0 \in A$ or $c_0 \in B$. In the first case we define $a_1 := c_0$ and $b_1 := b_0$, in the second $a_1 := a_0$ and $b_1 := c_0$. We then have $|b_1 - a_1| = 2^{-1}u$. Continuing this process recursively, we get two sequences $(a_n)$ and $(b_n)$ with the properties $|a_n - b_n| < 2^{-n}u$, $|a_n - a_{n+1}| \leq 2^{-n}u$, $|b_n - b_{n+1}| \leq 2^{-n}u$. Obviously the sequences converge to the same limit, say $c$. Now $c \in A \cup B$.

In case $c \in A$, then a tail of $(a_m)$ for sufficiently large $m$ is in an open interval around $c$ in $A$, and thus also a tail of $(b_m)$ for sufficiently large $m$ in an open interval around $c$ in $B$. So $c$ is in the intersection $A \cap B$.

Ditto for the case $c \in B$. $\qquad\square$

## 4.9  Higher Dimensions

Much of what is true of the continuum can be extended quite easily to the plane or space, or to higher dimensional spaces. We will not go into this systematically, but we show the reader that it is not so bad with "more variables".

For example, consider the plane, $\mathbb{R}^2$; the points are defined as pairs $(a_1, a_2)$ of real numbers. The distance of two points $a = (a_1, a_2)$ and $b = (b_1, b_2)$ is

$$\sqrt{(a_1 - b_1)^2 + (a_2 - b_2)^2}\,,$$

notation $\|a - b\|$. The role of intervals is now taken over by open circular disks: $O_a(r) = \{x \mid (\|a - x\| < r)\}$. The definition of the common terms now follows immediately. Here are some examples:

(1) $f : \mathbb{R}^2 \to \mathbb{R}^2$ is called *continuous at* $a$ if $\forall k \exists n (\|x - a\| < 2^{-n} \to \|f(x) - f(a)\| < 2^{-k})$.
(2) $A \subseteq \mathbb{R}^2$ is totally bounded if for every $k$ there is a finite number of points $p_0, p_1, \ldots, p_{n-1}$ with $\forall x \in A\ \exists m < n\ (\|x - p_m\| < 2^{-k})$.
(3) $A \subseteq \mathbb{R}^2$ is located if for every $a$ and $k < \ell$ $O_a(2^{-\ell}) \cap A = \emptyset$ or $O_a(2^{-k}) \cap A$ is inhabited.
(4) The distance from a point $a$ to a set $A$ is $d(a, A) = \inf\{\|a - x\| \mid x \in A\}$.

Many of the relevant theorems in this book can easily be generalized to $\mathbb{R}^n$.

# Chapter 5
# Functions and Continuity

**Abstract** We define continuity rigorously and show that continuous functions satisfy strong extensionality. Continuity, discontinuity, and differentiability are discussed. We find that classical gems like the Intermediate Value Theorem and Brouwer's Fixed Point Theorem are not constructively valid, as revealed by weak counterexamples. However, these theorems can be salvaged by proving approximate versions. In particular, using Sperner's Lemma, we do this for the Fixed Point Theorem in one and two dimensions.

## 5.1  The Concept of Continuity

Continuity is one of the most important concepts in science, and certainly in mathematics. A phenomenon behaves continuously if, popularly said, it does not make jumps. Before the advent of modern physics, especially quantum theory, the principle was generally accepted—"nature makes no leaps"— "Natura non facit saltus".[1] The idea is that things change gradually; it took millions of years for geological changes and the same goes for genetic processes. Continuity can also be observed in phenomena in everyday life. The heating of water in the boiler is continuous, a football describes a continuous arc from the tip of the forward's shoe to the hands of the goalkeeper, plants grow gradually, etc. We are sometimes tempted to say "she has progressed by leaps and bounds", but that is a manner of speaking.

Mathematics is the science par excellence that has made the concept of continuity precise. In fact, making concepts more precise is a very normal practice in science: there is some phenomenon that we observe, such as heat, speed, pressure, light, intelligence; after a while a scientific description or definition of the concept comes. For the concept of continuity, mathematics mainly focuses primarily on continuity of functions. In practice there are two equivalent ways to describe it.

---

[1] A principle appealed to by Leibniz, Linnaeus, and Darwin.



D. van Dalen et al., *Intuitionistic Analysis*, Springer Undergraduate Mathematics Series,
https://doi.org/10.1007/978-3-032-16491-9_5

**Fig. 5.1** Continuity in terms of intervals

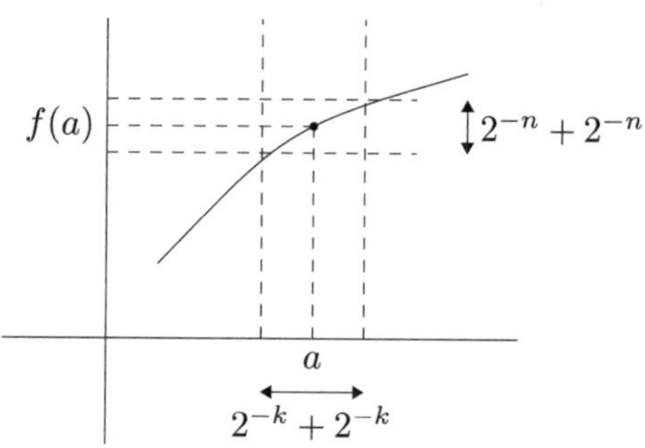



The first way is this. Consider a function $f$ at a point $a$. You want to control the change of values of $f$ near $f(a)$, i.e., you want to limit this variation around $f(a)$ to be, say, less than $2^{-n}$. This control is achieved when you can specify how much the values around $a$ are allowed to change so that the change around $f(a)$ remains within the desired limits. So we have to indicate an interval around $a$ so that, when the function is applied, the values of $f$ remain completely within the interval $(f(a) - 2^{-n}, f(a) + 2^{-n})$ (Fig. 5.1).

The exact definition of continuity of a real function is now that the following condition holds:

$$\forall n \exists k \forall x (|a - x| < 2^{-k} \rightarrow |f(a) - f(x)| < 2^{-n})$$

Because in mathematics the precise nature of the intervals is usually neglected, as a rule $\epsilon$ and $\delta$ are used for "small positive numbers". The condition will then look like this:

$$\forall \epsilon > 0 \; \exists \delta > 0 \; \forall x (|a - x| < \delta \rightarrow |f(a) - f(x)| < \epsilon)$$

The $\epsilon - \delta$ method is such a familiar mathematical tradition that $\epsilon$ and $\delta$ are automatically treated as small positive numbers without further specification.

The above means that continuity is a form of precision control. Think of some process that is controlled by the turning of a knob, for example setting the frequency of a receiver. What you want is that the result—visible on a dial—can be controlled well. If you want to get a fine-tuning for a certain frequency, then the pointer shouldn't suddenly jump away. You want to reach the setting slowly and precisely. That basically means that from a given position of the dial, we can make sure that the pointer does not move more than (say) 3 mm to the left or right by turning the knob no more than (say) 10° to the left or right. The millimeters then relate to $\epsilon$ and the degrees to $\delta$.

Slogan: continuity = unlimited precision.

The second description of continuity says "if you get very close to $a$, you also get very close to $f(a)$". The natural formulation here is

**Fig. 5.2** Continuity in terms of limits

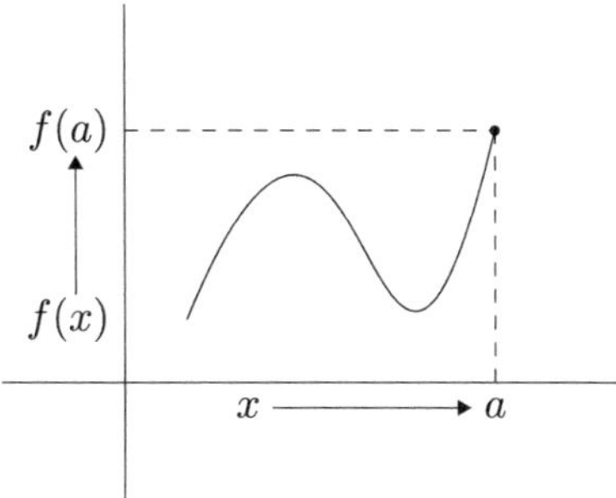


**Fig. 5.3** A discontinuous function

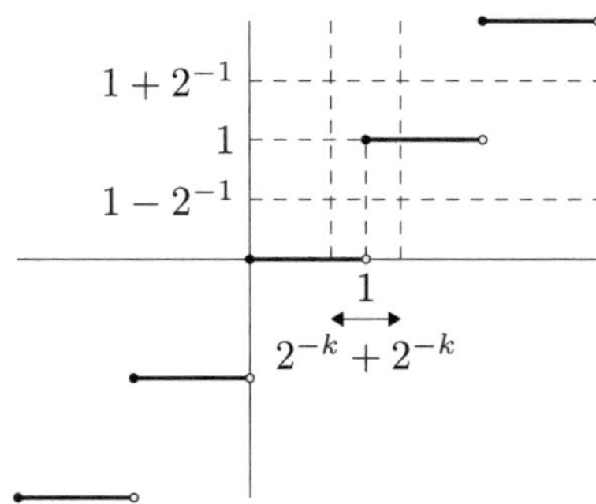


$$\lim_{x \to a} f(x) = f(a)$$

This formulation is easier to grasp, but keep in mind that you still have to use the definition of limit to apply it (Fig. 5.2).

Both definitions make it clear that no jumps can be made. As an illustration, we give a well-known discontinuous (i.e., non-continuous) function: $y = \text{int}(x)$, where $\text{int}(x)$ is the *integer* part of $x$ (Fig. 5.3). You're leaving out, so to speak, everything that comes after the decimal point in the decimal expansion of $x$. To be precise: $\text{int}(x)$ is the largest integer that is $\leq x$.

Now look at the value at 1. Although $\text{int}(1) = 1$, the limit of int does not exist at 1 at all (as $x$ approaches 1 from the left, $y$ never approaches 1). If we employ definition (1), then we want (for example) an interval $(1 - 2^{-k}, 1 + 2^{-k})$ in which $\text{int}(x)$ will lie within the interval $(1 - 1/2, 1 + 1/2)$. Unfortunately, the function int maps the desired small interval to two points, 0 and 1, and the point 0 falls outside the interval $(1 - 1/2, 1 + 1/2)$.

**Exercise 5.1** Prove that the functions $f$ and $g$ with

$$f(x) = \begin{cases} 0 \text{ if } x = 0 \\ 1/x \text{ if } x \neq 0 \end{cases} \qquad g(x) = \begin{cases} 0 \text{ if } x = 0 \\ x/|x| \text{ if } x \neq 0 \end{cases}$$

are discontinuous at 0.

We will come back to discontinuity in Sect. 6.6.


## 5.2   Continuous Functions


Continuous functions have a property that is completely trivial for classical mathematics, but which is not self-evident for intuitionistic mathematics. Classically and intuitionistically, for functions $f$ we have $a = b \to f(a) = f(b)$, and thus $f(a) \neq f(b) \to a \neq b$. Intuitionistically we would wish $f(a) \,\#\, f(b) \to a \,\#\, b$.

**Definition 5.2**  A function $f$ is called *strongly extensional* when

$$f(x) \,\#\, f(y) \to x \,\#\, y .$$

By the way, the contrapositive implication, $\neg x \,\#\, y \to \neg f(x) \,\#\, f(y)$, amounts to $x = y \to f(x) = f(y)$, which is the ordinary *extensionality* of functions.

The underlying argument for strong extensionality is that the apartness of the image points acts as a test for the apartness of the original ones. But there is actually no reason to believe that all functions are strongly extensional (for a weak counterexample in a specifically intuitionistic setting, see Theorem 6.22). This is different if we know that a function is continuous.

**Theorem 5.3**  *If $f$ is continuous, then $f(a) \,\#\, f(b) \to a \,\#\, b$.*

**Proof**  Let $f(a) \,\#\, f(b)$, then $|f(a) - f(b)| > 2^{-n}$ for a certain $n$. Now determine $p$ and $q$ such that

$$|a - x| < 2^{-p} \to |f(a) - f(x)| < 2^{-n-2} ,$$
$$|b - x| < 2^{-q} \to |f(b) - f(x)| < 2^{-n-2} .$$

Suppose there is an $x$ falling within both intervals $I_1 = (a - 2^{-p}, a + 2^{-p})$ and $I_2 = (b - 2^{-q}, b + 2^{-q})$, then we have

$$|f(a) - f(b)| \le |f(a) - f(x)| + |f(x) - f(b)| < 2^{-n-1} .$$

Hence, $I_1 \cap I_2 = \emptyset$.

Now we are going to determine rational intervals around $a$ and $b$: there are rational numbers $r_1, r_2, r_3, r_4$ such that $a - 2^{-p} < r_1 < a; a < r_2 < a + 2^{-p}; b - 2^{-q} < r_3 < b; b < r_4 < b + 2^{-q}$ (according to Theorem 4.11). It is clear that $(r_1, r_2) \cap (r_3, r_4) = \emptyset$; we then compare $r_2$ and $r_3$. If $r_2 < r_3$ then $a < b$ and if $r_2 > r_3$ then $a > b$. This shows that $a \,\#\, b$.                                                                          $\square$

Continuous functions are of great importance for all kinds of sciences. Much is known about these types of functions. We will now prove one of the most important theorems about continuous functions in classical mathematics. Next we will show that in intuitionistic mathematics the theorem is not universally valid.

**Fig. 5.4** Find a root between
$f(0)$ and $f(1)$, by classical
means

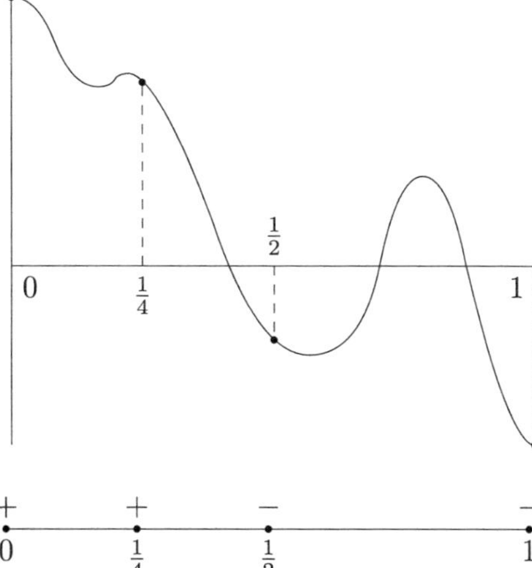



**Fig. 5.5**  Marking the signs
of $f(x)$ for $x$ that turned out
not be roots of $f$



**Theorem 5.4**  (Classical Intermediate Value Theorem) *If $f$ is continuous, $f(0) > 0$
and $f(1) < 0$, then there is a point between 0 and 1 where $f$ takes the value 0.*

First a heuristic consideration: put your pencil in the point $(0, 1)$ and then go without taking the pencil from the paper—after all, "continuous" means "no jumps"—to
the point $(1, -1)$, then you have to go over the $x$-axis along the way. And that yields
the requested point.

***Proof*** Using the example function in Fig. 5.4, we show the general reasoning. We
are going to try to find a point as desired. Determine $f(1/2)$, if this value is 0, then
we are done. If not, the value is positive or negative. We then mark $1/2$ with a + or a
−; there is now a situation where a − at distance $1/2$ is preceded by a +. That is the
starting point for the next step: determine the midpoint of this $+, -$-segment. If the
value of $f$ at that point is 0, then we are done, otherwise we can mark this midpoint
with a + or a − again (Fig. 5.5).

We get another segment with a $+, -$, now of length $2^{-2}$. It is now clear what to
do: we successively choose new midpoints, which gradually yield smaller segments
with a $+, -$. At step $n$ we get a segment of length $2^{-n}$.

Call the found midpoints $a_1, a_2, a_3, \ldots$. It is clear that the sequence converges
(each point $a_n$ is less than $2^{-n}$ from all succeeding points). So the sequence has
a limit, say $a$. If $f(a) = 0$, we are done. If not, then $f(a)$ is positive or negative.
Consider $f(a) > 0$, then there is a $k$ with $f(a) > 2^{-k}$. Because of the continuity of
$f$ there is an $\ell$ such that $|x - a| < 2^{-\ell} \rightarrow |f(x) - f(a)| < 2^{-k}$.

So all numbers closer than $2^{-\ell}$ to $a$ get positive values under $f$ (see Fig. 5.6).
Now consider that the point $a_n$ is always at a distance less than $2^{-n}$ from a point
marked "−", say $b_n$, then we see that $a$ is also a limit of such points marked "−".

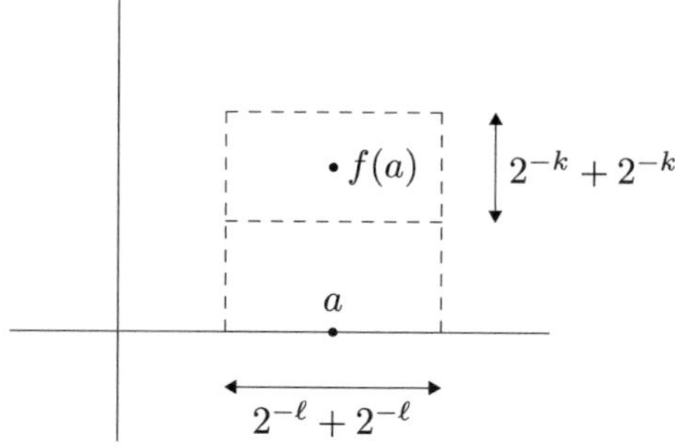

**Fig. 5.6** For points close to $a$, $f(a)$ is positive



More precisely: because $\lim(a_n) = a$, there is an $a_n$ with $|a_n - a| < 2^{-\ell-1}$, and we may choose $n$ greater than $\ell + 1$. Now $|b_n - a| \le |b_n - a_n| + |a_n - a| = 2^{-\ell-1} + 2^{-\ell-1} = 2^{-\ell}$, so $f(b_n) > 0$. Contradiction.

In exactly the same way we see that $f(a) < 0$ is impossible. *Conclusion*: $f(a) = 0$.
$\qquad\qquad\qquad\qquad\qquad\qquad\qquad\qquad\qquad\qquad\qquad\qquad\qquad\qquad\qquad\qquad\square$

The question is: in this proof we described what looked like a procedure, but does it really consist of steps that can each be carried out? The difficulty in the procedure lies in assigning a $+$, $-$, or a $0$ to the midpoints.

After all, even if we know how the value of $f$ at such a point is defined, to determine whether its value is positive, zero, or negative, we need the law of trichotomy. And we have seen (Sect. 4.5) that that law is constructively not correct. This is where things immediately go wrong. Also, at the end we applied proof by contradiction. Therefore, the above proof is not acceptable to the intuitionist.

Yet despite all that, the theorem could still be correct; it could just be a lack of cleverness on our part that prevents us from finding an acceptable proof. However, this is not the case, as we can give a weak counterexample.

Define a piecewise linear function $f$ from $[0, 3]$ to $\mathbb{R}$ by declaring that it first passes from $(0, 1)$ to $(1, a)$, from there to $(2, a)$, and from there to $(3, -1)$. Here $a$ is one of those values for which, as yet, we cannot determine whether $a \ge 0$ or $\le 0$ (see Sect. 4.5); of course, to draw a graph with $a$ on the $y$-axis we have to make a choice, so Fig. 5.7 gives just a rough idea of the real situation. The function satisfies

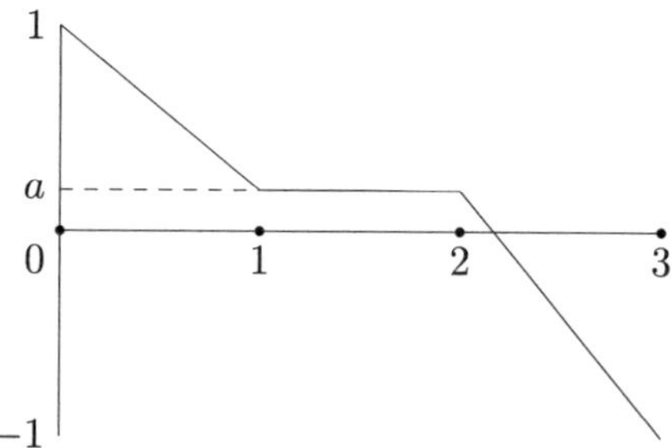

**Fig. 5.7** The graph of $f$ if we knew that $a$ is positive

the condition: $f(0) > 0$, $f(3) < 0$. Now suppose that we can construct a $b$ with $f(b) = 0$, then $b$ must be to the right of 1 or to the left of 2. This is a consequence of Theorem 4.14 ($x < y \rightarrow x < z \vee z < y$). But if $b > 1$ then $a \geq 0$, and if $b < 2$ then $a \leq 0$.

So if the Intermediate Value Theorem is correct, then by constructing enough of $(b_n)$ we could decide for an arbitrary $a$ whether $a \leq 0$ or $a \geq 0$. That is not the case. So the Intermediate Value Theorem is not correct. The standard trick again: reduce the present problem to an as yet undecidable statement.

We therefore cannot accept the Intermediate Value Theorem in its entirety. However, there is still something to salvage. The difficulties are caused by the occurrence of almost horizontal pieces. If we ban those, we can still achieve something, and obtain an intuitionistic version. So here it is again, with a condition that ensures that no near-horizontal pieces occur locally in the graph.

**Theorem 5.5**  *If $x \mathbin{\#} y \rightarrow f(x) \mathbin{\#} f(y)$ and $f(0) > 0$ and $f(1) < 0$, then $f$ has a zero between 0 and 1.*

**Proof**  The proof is similar to the classical proof, but contains a few refinements. This time we are going to perform repeated *threefold* divisions.

We determine the values at $1/3$ and $2/3$. Since $1/3 \mathbin{\#} 2/3$, by the hypothesis of the theorem $f(1/3) \mathbin{\#} f(2/3)$. By property (2) of apartness (Theorem 4.31), we get $f(1/3) \mathbin{\#} 0 \vee f(2/3) \mathbin{\#} 0$. If $f(1/3) \mathbin{\#} 0$, then $f(1/3) > 0$ or $f(1/3) < 0$, and similarly for $f(2/3) \mathbin{\#} 0$. So if we mark points in the domain as we did in the proof of Theorem 5.4, by hypothesis of the theorem 0 is marked $+$ and 1 is marked $-$, and by our assumption at least one of the two newly chosen domain points is going to be marked with a $+$ or a $-$.

As before, we're looking for a "$+$"-point followed by a "$-$"-point. If we are lucky the distance is $1/3$, and if we are unlucky the distance is $2/3$. In a proof you always have to take the worst into account, which means that we can assume that we find a "$+, -$"-segment with length $2/3$.

We will give the reasoning for the case that $1/3$ is marked with a $+$, and hence the segment we find is $(1/3, 1)$. Now we repeat the threefold division—and look at domain points $5/9$ and $7/9$. At least one of the values $f(5/9)$, $f(7/9)$ is apart from 0; therefore one of these values is positive or negative.

Looking for the next (smaller) "$+, -$"-segment, we continue to consider what might be the longest segments to occur. Here is a list of the possibilities:

$$
\begin{array}{cccc}
\frac{1}{3} & \frac{5}{9} & \frac{7}{9} & 1 \\
\hline
+ & ? & + & - \\
+ & ? & - & - \\
+ & + & ? & - \\
+ & - & ? & - \\
\end{array}
$$

The reason for the question marks is this. It is quite possible that for both $f(5/9)$ and $f(7/9)$ the calculated values are positive or negative, but we have to keep in mind

**Fig. 5.8** Distant points with
the same function value

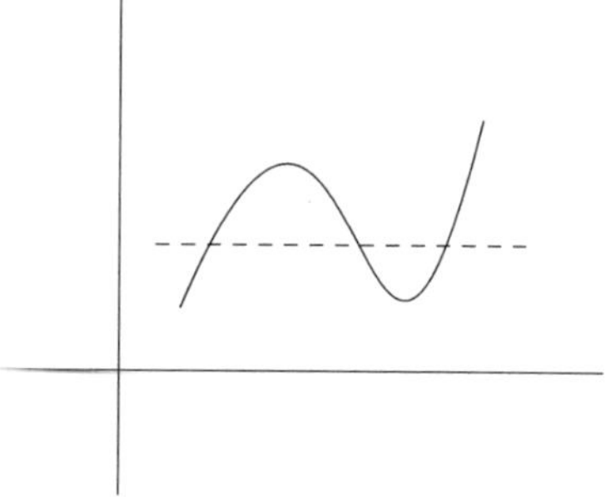

that one of those two values is close to 0, so perhaps we don't know if it is 0, positive or negative. So in keeping with our worst case scenario, we do not take those places in the table into account.

In that case the second and third rows yield the largest segments, with length $(2/3)^2 = 4/9$. We see that the length of the "+, $-$"-segments we are seeking becomes $2/3$ of what it was.

So we continue the procedure of making triples, generating a sequence of "+"-points $a_1, a_2, a_3, \ldots$ and a corresponding sequence of adjacent "$-$"-points, $b_1, b_2, b_3, \ldots$. And since the sequence $((2/3)^n)$ has limit 0, the sequence of "+"-points converges to a point $a$.

Claim: $f(a) = 0$. Suppose $f(a) \mathbin{\#} 0$, then $f(a) > 0$ or $f(a) < 0$. Assume that $f(a) > 0$. Then we can use the continuity of $f$ exactly as in the classical proof to get a contradiction. Entirely analogously we see that $f(a) < 0$ is impossible, so $\neg f(a) \mathbin{\#} 0$. That is equivalent to $f(a) = 0$. $\qquad\square$

At first glance, the additional condition is not much of a limitation, but in practice only monotone functions appear to satisfy it, i.e., functions that keep or reverse the order:

**Definition 5.6** (*Monotonicity*) $f$ is

$$
\begin{aligned}
&\textit{monotone increasing} &&\text{if } x < y \to f(x) < f(y), \\
&\textit{monotone decreasing} &&\text{if } x < y \to f(x) > f(y), \\
&\textit{monotone non-decreasing} &&\text{if } x < y \to f(x) \le f(y), \\
&\textit{monotone non-increasing} &&\text{if } x < y \to f(x) \ge f(y).
\end{aligned}
$$

Obviously, if the graph goes up and down you can find two distant points with the same value (Fig. 5.8). Therefore, the following statement is not so surprising.

**Theorem 5.7** *If a continuous function $f$ satisfies $x \mathbin{\#} y \to f(x) \mathbin{\#} f(y)$, then $f$ is monotone increasing or monotone decreasing.*

**Proof** Take two points $a$ and $b$ apart from each other, then $a < b$ or $a > b$. Consider the case $a < b$. Now determine $f(a)$ and $f(b)$. By assumption, $f(a) \mathbin{\#} f(b)$. There are now two possibilities: $f(a) < f(b)$ or $f(a) > f(b)$.

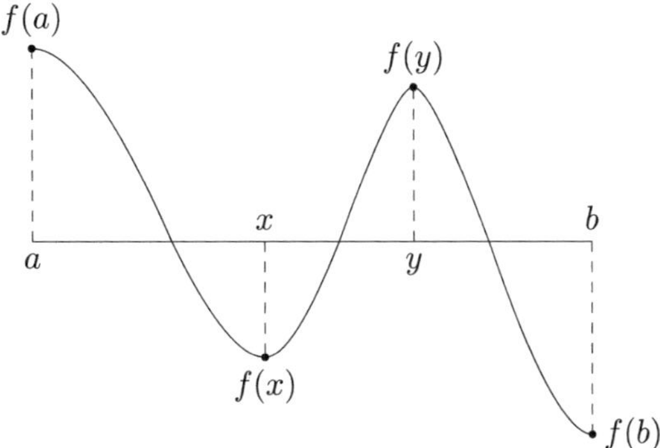

**Fig. 5.9** Suppose there exist such $x$ and $y$

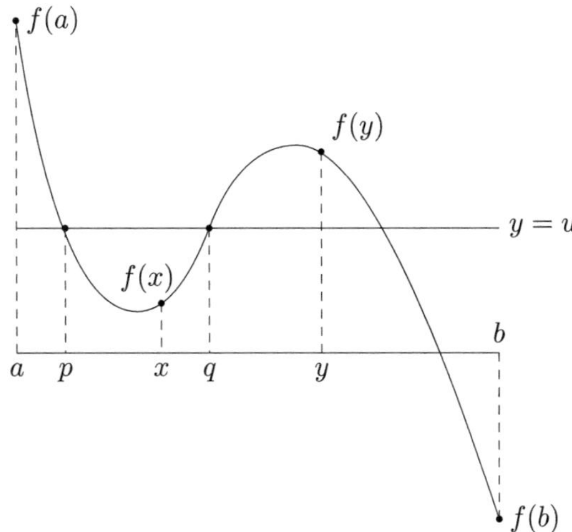

**Fig. 5.10** A case where $f(a) > f(y)$

Let us take the case $f(a) > f(b)$. We have to show that $f$ is monotone decreasing. To begin with, we show that $f$ decreases monotonically between $a$ and $b$. Once we have done that, the monotonicity for the entire real line is not difficult to establish.

Suppose there are points $x$ and $y$ strictly between $a$ and $b$ such that $x < y$ and $f(x) < f(y)$. Figure 5.9 shows a possible case.

Since $x \ne y$, either $a \ne x$ or $a \ne y$. Consider the case $a \ne y$. By assumption, $f(a) \ne f(y)$. There are two possibilities: $f(a) > f(y)$, or $f(a) < f(y)$.

(i) First consider $f(a) > f(y)$ (Fig. 5.10).

Take the average of $f(x)$ and $f(y)$, say $u$. According to the previous theorem, the line $y = u$ intersects the graph of $f$ between $a$ and $x$ and between $x$ and $y$, say in $p$ and $q$ (why?). You see that $p \ne q$ (how?). Therefore, $f(p) \ne f(q)$ holds, but that is not possible, because $f(p) = f(q) = u$.

(ii) Case two: $f(a) < f(y)$ (Fig. 5.11).

Find the mean of $f(a)$ and $f(y)$, say $v$. Now draw the line $y = v$. From $f(a) < f(y)$, it follows that $f(a) < v$, and further that $f(y) > v > f(a) > f(b)$. So there

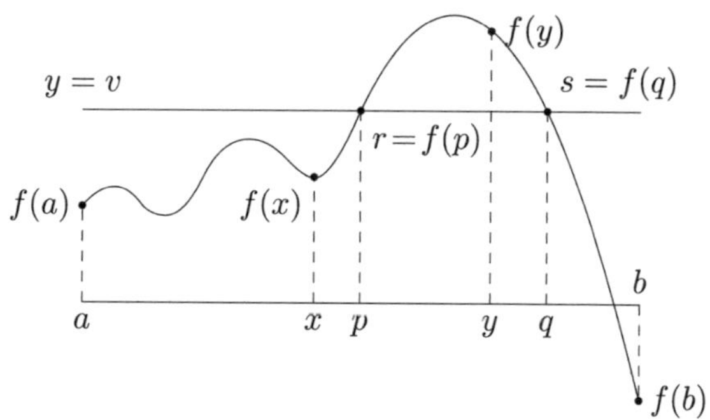

**Fig. 5.11** A case where $f(a) < f(y)$



are intersections $r$ and $s$ of the graph with the line $y = v$ between $f(a)$ and $f(y)$ and between $f(y)$ and $f(b)$. However, that is impossible (why?). This shows that it cannot happen that $f(x) < f(y)$; so $f(x) > f(y)$. *Conclusion*: $f$ is monotone decreasing between $a$ and $b$.

We now consider the points outside the interval $(a, b)$. Suppose there are two points $x$ and $y$ with $x < y$ and $f(x) < f(y)$. Then we choose two points $c$ and $d$ so that $c$ is to the left of $a, b, x$, and $y$, and $d$ is to the right of $a, b, x$, and $y$. On this interval $(c, d)$ $f$ must be monotone decreasing, by the same argument which showed that $f$ is monotone decreasing on $(a, b)$. But that is contrary to $f(x) < f(y)$. *Conclusion*: $f(x) > f(y)$, and so $f$ is decreasing everywhere.  $\square$

The weak counterexample to the unrestricted Intermediate Value Theorem suggests that the problems are caused by pieces of the function that are nearly horizontal near the $x$-axis. The following theorem includes a condition that forbids this phenomenon.

**Theorem 5.8** *If $f$ is a continuous function on $[a, b]$ with $f(a)$ positive and $f(b)$ negative, which also satisfies $\forall x \forall y (x < y \rightarrow \exists z(x < z < y \wedge f(z) \# 0))$, then $f$ has a zero between $a$ and $b$.*

**Exercise 5.9** Prove Theorem 5.8 using an approximation (again using triples).

After all these reinforcements of the assumptions (and weakening of the theorem), it would be interesting to know if there is anything left of the old version at all. The answer is (fairly) positive:

**Theorem 5.10** (approximate Intermediate Value Theorem) *If $f$ is continuous on $[a, b]$ and $a < b$, $f(a) > 0$, $f(b) < 0$ then $\forall k \exists x |f(x)| < 2^{-k}$.*

We will give two proofs.

1. We construct a direct approximation of the desired point. Let $k$ be given. Suppose

$$a_0 = a$$
$$b_0 = b$$
$$c_0 = (b + a)/2.$$

With $-2^{-k-1} < 2^{-k-1}$ and Theorem 4.14 we have

$$f(c_0) > -2^{-k-1} \vee f(c_0) < 2^{-k-1}$$

If the left disjunct is true, define

$$a_1 = c_0, b_1 = b_0 \, ,$$

and if the right one is,

$$a_1 = a_0, b_1 = c_0 \, .$$

(If both are true, you choose.) We continue this process. Suppose $c_{n-1} = (b_{n-1} + a_{n-1})/2$, then

$$f(c_{n-1}) > -2^{-k-1} \vee f(c_{n-1}) < 2^{-k-1}$$

If the left disjunct is true, define

$$a_n = c_{n-1}, b_n = b_{n-1} \, ,$$

and if the right one is,

$$a_n = a_{n-1}, b_n = c_{n-1} \, .$$

(If both are true, you choose.) The result is two sequences $(a_n)$, $(b_n)$ with $\forall n |a_n - b_n| < 2^{-n}|a - b|$.

Check that both sequences are Cauchy sequences, and that they determine the same number $w$.

Prove that $-2^{-k} < f(w) < 2^{-k}$. $\qquad\square$

The theorem also has a smart proof without approximation (at least not a direct one).

2. The sets $\{x \mid f(x) > -2^{-n}\}$ and $\{x \mid f(x) < 2^{-n}\}$ are open and cover $[a, b]$. Now apply Theorem 4.75. $\qquad\square$

Zeros of functions are directly related to fixed points of functions. A fixed point of $f$ is an element $x$ in its domain for which $f(x) = x$. Now if $f : [0, 1] \to [0, 1]$ is a continuous function, then a fixed point of $f$ is a zero of $g$ with $g(x) = f(x) - x$: $f(x) = x \leftrightarrow g(x) = 0$.

A weak counterexample analogous to that given in Fig. 5.7 shows that a continuous map of $[0, 1]$ in itself need not have a fixed point. We will come back to this and related matters in Sect. 5.6.

## 5.3  Discontinuity

A function is discontinuous at a point $a$ if the "encircling technique" of continuity does not work. That is, for an arbitrarily small neighborhood $(f(a) - 2^{-k}, f(a) + 2^{-k})$, you can't always find a small neighborhood $(a - 2^{-n}, a + 2^{-n})$ that is mapped

$$f(x) = \begin{cases} 0 & \text{if } x \leq 0 \\ 1 & \text{if } x > 0 \end{cases}$$

**Fig. 5.12** A discontinuity at 0

entirely into the first interval. At the end of Sect 5.1 we have already seen an example of such a discontinuous function.

We give another example. See Fig. 5.12. At 0 the function is discontinuous, as you can immediately see (perhaps most easily with the limit definition: $\lim_{x \to 0}$ does not exist). However, there is a catch: the function isn't defined for all real numbers (in other words, not on the whole $x$-axis). The difficulty is caused by the numbers that can't tell whether they are to the left of 0, to the right of 0, or at 0. Take one of the Brouwerian oscillating sequences, which determines a number $a$ for which $a < 0 \vee a = 0 \vee a > 0$ cannot be decided. We can't compute the function value for such $a$. So $f(a)$ is undefined. That situation doesn't exist with the functions we usually consider, say $\sin(x)$ or $x^3 - 2x^2 + 5$. Those functions have no gaps in their domain of definition. You might now suspect that functions that are defined everywhere (so-called *total functions*) can have no discontinuities. Brouwer proved in 1924—to the astonishment of his classical colleagues—that all total real functions are continuous. See Theorem 6.10.

## 5.4   Uniform Continuity

The function $f(x) = 1/x$ is continuous on the open interval $(0, 1)$, as can be easily checked. Its graph in Fig. 5.13 shows that, for a fixed $\delta$, the $\epsilon$ get larger and larger the closer we get to 0.

Conversely, if we fix $\epsilon$ and move to the left, the length of the corresponding $\delta$ drops below any boundary. There is, so to speak, no "one size fits all". But if the function does not go off to infinity, we can do with a single estimate for $\delta$. This phenomenon is called *uniform continuity*.

**Definition 5.11** We call a function $f$ on a closed segment $[a, b]$ *uniformly continuous* if

$$\forall k \exists \ell \, \forall x_1, x_2 \in [a, b] \, (|x_1 - x_2| < 2^{-\ell} \to |f(x_1) - f(x_2)| < 2^{-k})$$

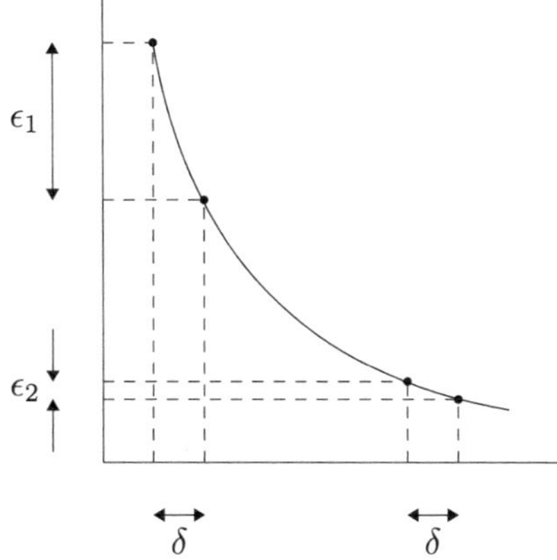


**Fig. 5.13**  A decreasing curve

In the $\epsilon$-$\delta$ formulation we get

$$\forall \epsilon > 0 \; \exists \delta > 0 \; \forall x_1, x_2 \in [a, b] \; (|x_1 - x_2| < \delta \to |f(x_1) - f(x_2)| < \epsilon)$$

One of the most important theorems in elementary analysis says "a continuous function on a closed segment is uniformly continuous". The proof is based on the Heine-Borel Theorem.

**Theorem 5.12** (Heine-Borel) *If $[a, b]$ is covered by a set of open intervals, then it is already covered by a finite subset.*

***Proof*** (*Classical*) For convenience, we consider $[0, 1]$. There is a set $\mathcal{A}$ of open intervals of $[0, 1]$ in the sense of the topology of $[0, 1]$ covering $[0, 1]$.[2]

Call *a reachable* if $[0, a]$ is covered by finitely many elements of $\mathcal{A}$. The set $X$ of reachable numbers is bounded and non-empty. According to classical analysis, $X$ has a supremum $s$. Assume $s < 1$. Then there is an interval $(a, b) \in \mathcal{A}$ with $a < s < b$. Take $p$ and $q$ such that $a < p < s < q < b$. Now $p$ is reachable, that is, there is a finite cover of $[0, p]$. Adding $(a, b)$ to this yields a finite cover of $[0, q]$. Contradiction. So $s = 1$, and $[0, 1]$ has a finite subcover of $\mathcal{A}$. $\qquad\square$

The proof of this theorem makes essential use of PEM, so we cannot simply copy it in constructive analysis. The situation is somewhat complicated, because within a certain approach to constructive mathematics—algorithmic or *recursive analysis*—a weak counterexample to Heine-Borel can be given, while in intuitionism, with a different understanding of real numbers and different principles for them, the theorem can be proved. See Sect. 6.4 and in particular Theorem 6.16. So Heine-Borel is one of those "independent" principles, which are sometimes acceptable and sometimes not acceptable.

---

[2] That is, we consider the "relative topology" of the interval $[0, 1]$ induced by the usual Euclidean topology on the real numbers; in the  relative topology, the open sets are the intersections of the open sets in the  Euclidean topology with $[0, 1]$. In particular, then $[0, a)$ and $(a, 1]$ are also open sets.

Since uniformly continuous functions have many advantages in a constructive treatment of analysis, there is a pragmatic way out: make use of uniformly continuous functions, whenever possible. Below is a theorem that provides us with many real functions.

**Theorem 5.13** *If $f$ is a uniformly continuous function on the rational interval $[a, b]$, then there is exactly one uniformly continuous extension of $f$ to the real interval $[a, b]$.*

***Proof***

(1) We define the extension via Cauchy sequences. A real number $w \in [a, b]$ can be given by a Cauchy sequence of rational numbers that are all in $[a, b]$. To see this, take $w \geq a$ in $[a, b]$, and let $w = (w_n)$. Now define

$$c_n = \begin{cases} a, & \text{if} \quad w_n < a \\ w_n, & \text{if} \quad a \leq w_n \leq b \\ b, & \text{if} \quad b < w_n. \end{cases}$$

Then $w = c$, with $c$ represented by a Cauchy sequence with rational terms in $[a, b]$.

For such a $c$ we define $f^*(c) := (f(c_n))$. Show that $f$ transfers rational Cauchy sequences into Cauchy sequences. Obviously $f^*$ is defined for real numbers on $[a, b]$.

(2) We have yet to show that equivalence is preserved: $(c_n) \sim (d_n) \to (f(c_n)) \sim (f(d_n))$. We have to prove $\forall k \exists n \forall m > n(|f(c_m) - f(d_m)|) < 2^{-k}$; choose a $k$ and determine the $\ell$ that exists according to the uniform continuity: $|x - y| < 2^{-\ell} \to |f(x) - f(y)| < 2^{-k}$. Then determine an $n$ such that $\forall m > n|c_m - d_m| < 2^{-\ell}$, then also $\forall m > n(|f(c_m) - f(d_m)| < 2^{-k})$. So the function $f^*$ is well-defined.

(3) We must now show that $f^*$ is uniformly continuous on $[a, b]$. To this end, determine the $\ell$ on the rational segment at $k$ such that

$$|x - y| < 2^{-\ell} \to |f(x) - f(y)| < 2^{-k-1}.$$

Consider real $c$ and $d$ such that $|c - d| < 2^{-\ell-2}$, and determine

$$n_1 \text{ such that } \forall m > n_1|c - c_m| < 2^{-\ell-2},$$
$$n_2 \qquad \forall m > n_2|d - d_m| < 2^{-\ell-2}.$$

Let $n = \max(n_1, n_2)$. Then, for $m > n$,

$$|c_m - d_m| = |c_m - c + c - d + d - d_m|,$$
$$< |c_m - c| + |c - d| + |d - d_m|,$$
$$< 2^{-\ell-2} + 2^{-\ell-2} + 2^{-\ell-2} < 2^{-\ell}.$$

So $\forall m > n(|f(c_m) - f(d_m)| < 2^{-k-1})$. Now determine

$$n_3 \text{ such that } \forall m > n_3(|f^*(c) - f(c_m)| < 2^{-k-2}),$$
$$n_4 \qquad\qquad \forall m > n_4(|f^*(d) - f(d_m)| < 2^{-k-2}).$$

If $n' = \max(n, n_3, n_4)$, then for $m > n'$

$$|f^*(c) - f^*(d)| = |f^*(c) - f(c_m) + f(c_m) - f(d_m) + f(d_m) - f^*(d)|,$$
$$< |f^*(c) - f(c_m)| + |f(c_m) - f(d_m)| + |f(d_m) - f^*(d)|,$$
$$< 2^{-k-2} + 2^{-k-1} + 2^{-k-2} = 2^{-k}.$$

This demonstrates the uniform continuity of $f^*$. $\qquad\qquad\qquad\qquad\qquad\square$

A somewhat cleverer treatment even shows that the $\ell$ and $k$ can be preserved (see [83, p. 261])—preserving the modulus of continuity.

The foregoing theorem allows us to "paste functions together". Without extra precautions, functions as used in the weak counterexample to the Intermediate Value Theorem are problematic, because they are defined by cases: "if $x < \cdots$ then $f(x) = \cdots$, and if $x \geq \cdots$ then $f(x) = \cdots$". Such a case distinction is in general not constructive. The paste method is a useful tool for creating new functions.

In classical mathematics continuous functions on closed intervals have a maximum and a minimum. This generally does not apply here. Two examples:

(1) Consider the linear function $f(x) = ax$ on $[0, 1]$, where $a$ is one of those numbers for which it is unknown whether $a \leq 0$ or $0 \leq a$. Suppose that $f$ has a maximum at the point $t$. Now $t \# 0 \vee t \# 1$; if $t \# 0$ then $0 \leq a$ and if $t \# 1$, then $a \leq 0$. So we cannot claim that $f$ has a maximum.
(2) The polynomial $-3x^4 + 4ax^3 + 6x^2 - 12ax$, where $a$ is a number whose location relative to 0 is unknown, has a supremum $3 + 8|a|$, but no maximum. Show that we cannot make a choice between the argument values $-1$ and $+1$. This example is given by Heyting [47, Sect. 3.4.3].

But we do have the following:

**Theorem 5.14** *If $f$ is uniformly continuous on $[a, b]$, then $f$ has a* sup *and an* inf.

***Proof*** We show that $\operatorname{ran}(f)$ is totally bounded; recall Definition 4.73. For given $k$, determine an $\ell$ such that $|x_0 - x_1| < 2^{-\ell} \to |f(x_0) - f(x_1)| < 2^{-k}$. Now divide $[a, b]$ into equal pieces $[c_i, c_{i+1}]$ of length $u$, where $u \leq 2^{-\ell}$. Check that for every $x \in [a, b]$ a division point $c_i$ can be found with $|x - c_i| < 2^{-\ell}$. Then it also holds that every point $f(x)$ in $\operatorname{ran}(f)$ has a distance smaller than $2^{-k}$ from one of the points $f(c_i)$. This shows that $\operatorname{ran}(f)$ is totally bounded. According to Theorem 4.74, $\operatorname{ran}(f)$ is now also located. And therefore, by Theorem 4.69, $\operatorname{ran}(f)$ has a sup and an inf. $\qquad\square$

## 5.5  Differentiability

We will deal with differentiability on closed intervals. It makes sense to use uniformity here also.

**Definition 5.15**  A function $f$ is called *uniformly differentiable* on $[a, b]$ with $a < b$ if there is a function $f'$ and a *modulus function* $\alpha$ with

$$\forall k \; \forall x \in [a, b] \; \forall y \in [a, b]$$
$$(|x - y| < 2^{-\alpha(k)}$$
$$\rightarrow$$
$$|f'(x)(y - x) - (f(y) - f(x))| < 2^{-k}|y - x|)$$

The function $f'$ is called the *derivative* of $f$.

The geometric meaning is clear: $f'(x)$ indicates the slope at the point $f(x)$; the modulus function tells you how close to $x$ you have to pick the point $y$ to get a deviation smaller than $2^{-k}$.

We note first that differentiability implies continuity.

**Theorem 5.16**  *A uniformly differentiable function is uniformly continuous, and its derivative, $f'(x)$, is also uniformly continuous.*

This theorem is a small variation on a theorem from classical analysis.

**Exercise 5.17**  Prove Theorem 5.16.

The usual properties can be derived, e.g.,

$$(f + g)'(x) = f'(x) + g'(x)$$
$$(f \cdot g)'(x) = f'(x) \cdot g(x) + f(x) \cdot g'(x)$$
$$(af)'(x) = af'(x)$$
$$\left(\frac{1}{f}\right)'(x) = -\frac{f'(x)}{f(x)^2} , \text{ where } \inf(f) > 0 \text{ on } [a, b]$$

In classical analysis, Rolle's Theorem states that if a differentiable function has the same value in two points, the derivative between those two points must have a zero. In this form the theorem is not constructively valid, but there is an approximate Rolle's Theorem, which we will simply call *Rolle's Theorem* here.

**Theorem 5.18**  (Rolle's Theorem) *If $f$ is uniformly differentiable on $[a, b]$, $a < b$, and $f(a) = f(b)$, then $\forall n \; \exists x \in [a, b] \; (|f'(x)| < 2^{-n})$, i.e., $\inf(|f'|) = 0$ on $[a, b]$.*

***Proof***  Set $m = \inf\{|f'| \mid x \in [a, b]\}$ (see Theorem 5.14). Assume $m > 0$, then there is a $k_1$ such that $m > 2^{-k_1}$. From $|f'(a)| \geq m$ follows $f'(a) \leq -m \vee f'(a) \geq m$. Consider the case $f'(a) \geq m$; the other case is analogous. Suppose that for a

$c \in [a, b]$ it is the case that $f'(c) < m$. Then by our assumption that $m > 0$ and the definition of $m$, we must have $f'(c) < 0$. Now if $f'(a) \geq m > 0$ and $f'(c) < 0$, and $f'$ is uniformly continuous by Theorem 5.16, we apply the (approximate) Intermediate Value Theorem (Theorem 5.10) and get a $y \in [a, c]$ with $|f'(y)| < 2^{-k_1} < m$. This is contrary to the fact that $m$ above is given as the infimum.

*Conclusion*: $\forall x \in [a, b]\ (f'(x) \geq m).$[3]

Now apply the uniform differentiability of $f$:

$$\exists k (|x - y| < 2^{-k} \rightarrow |f'(x)(y - x) - (f(y) - f(x))| \leq 2^{-1} m |y - x|.$$

We now divide the interval into pieces smaller than $2^{-k}$: determine a $p$ such that $(b - a)/p < 2^{-k}$. The distribution points are $d_i = a + i(b - a)/p, i \leq p$. Thus,

$$
\begin{aligned}
0 &= f(b) - f(a) \\
&= \sum_{i=0}^{p-1} (f(d_{i+1}) - f(d_i)) \\
&= \sum_{i=0}^{p-1} f'(d_i)(d_{i+1} - d_i) + \sum_{i=0}^{p-1} (f(d_{i+1}) - f(d_i) - f'(d_i)(d_{i+1} - d_i)) \\
&\geq \sum_{i=0}^{p-1} m(d_{i+1} - d_i) - \sum_{i=0}^{p-1} \frac{1}{2} m(d_{i+1} - d_i) = \frac{1}{2} m(b - a) > 0.
\end{aligned}
$$

Contradiction. So $m = 0$. $\qquad\square$

The following approximation immediately follows from Rolle's Theorem.

**Corollary 5.19** *If $f$ is uniformly differentiable on $[a, b]$, $a < b$, then*

$$\forall k\ \exists x \in [a, b]\ |f(b) - f(a) - f'(x)(b - a)| \leq 2^{-k}.$$

***Proof*** Define $g$ by $g(x) := (x - a)(f(b) - f(a)) - f(x)(b - a)$. Note that $g(a) = -f(a)(b - a) = g(b)$. Now apply Rolle: $\forall k\ \exists x \in [a, b]\ (|g'(x)| \leq 2^{-k})$, so $|f(b) - f(a) - f'(x)(b - a)| \leq 2^{-k}$ for the $x$ found. $\qquad\square$

In the standard textbooks this corollary is usually found in the form: there is a $\theta$ between 0 and 1 such that $f(a + h) = f(a) + f'(a + \theta h)$. This identity cannot be demonstrated constructively; for us there is just the approximative version.

**Lemma 5.20** *For a uniformly differentiable function $f$ on $[a, b]$,*

$$\forall c \in [a, b]\ \left(f'(c)\ \#\ 0 \rightarrow \forall k\ \exists x\ (|c - x| < 2^{-k} \wedge f(x)\ \#\ f(c))\right).$$

---

[3] We are grateful to a referee for replacing our original argument up to this point with a simpler one.

In words: if the derivative of such a function at some point is apart from 0, then as close to that point as we wish, we can calculate a point at which the function value is apart from that at the first. That is, the function is far from flat in that area.

***Proof*** For given $c$, we have $a < c \vee c < b$ (Theorem 4.14). Here we treat the first case. Assume that $f'(c) \mathop{\#} 0$, and let $k$ be given. The seeming difficulty is that our knowledge is about intervals, but we are asked for a point. Corollary 5.19 above is of the right kind to help us out. We begin by determining an interval that contains $c$ and in which the derivative is guaranteed to be close enough to $f'(c)$ to be likewise apart from 0. To this end, we take "close enough" to mean: differing by less than half.

(1)  Choose an $l$ such that $2^{-l} < 2^{-1}|f'(c)|$. By Theorem 5.16, $f'$ is uniformly continuous. So from $l$ we can determine an $m$ such that

$$\forall z \in [c - 2^{-m}, c]\ |f'(c) - f'(z)| < 2^{-l} ,$$

and therefore

$$\forall z \in [c - 2^{-m}, c]\ |f'(z)| > 2^{-1}|f'(c)|.$$

(2)  Our plan is to show that $f(c - 2^{-m})$ is apart from $f(c)$, by showing that the straight line connecting the two is not horizontal, and that we can say something positive about the extent to which it is not. In this step, we do the first.

By Corollary 5.19,

$$\forall n\ \exists y \in [c - 2^{-m}, c]\ |f(c) - f(c - 2^{-m}) - f'(y)2^{-m}| \leq 2^{-n} ,$$

and so, dividing by $2^{-m}$,

$$\forall n\ \exists y \in [c - 2^{-m}, c]\ (|(f(c) - f(c - 2^{-m}))/2^{-m} - f'(y)| \leq 2^{m-n}) .$$

This means that for any given $n$ there is a $y$ in the interval such that the slope of the straight line between the end points of the interval is within $2^{m-n}$ of $f'(y)$. Since this holds for arbitrarily large $n$, and, by step (1), for all $y$ that we find we have $|f'(y)| > 2^{-1}|f'(c)|$, that connecting line is not horizontal. (Recall that we have assumed that $f'(c) \mathop{\#} 0$.)

(3)  To establish apartness, we must go further and determine a lower bound on the difference between $f(c)$ and $f(c - 2^{-m})$.
  We write $S$ for $(f(c) - f(c - 2^{-m}))/2^{-m}$.
  Recall the triangle inequality (Theorem 4.38): $|p + q| \leq |p| + |q|$. Applying this to $p = f'(y) - S$ and $q = S$ yields $|f'(y)| \leq |f'(y) - S| + |S|$ and then, using $|f'(y) - S| = |S - f'(y)|$, we get $|f'(y)| - |S - f'(y)| \leq |S|$. With the inequalities we found in step (2), now $2^{-1}|f'(c)| - 2^{m-n} \leq |S|$.

As is also clear from step (2), this holds for any $n$, so we choose one such that $2^{m-n} < 2^{-1}|f'(c)|$. Then $0 < |S|$, which implies $S \mathop{\#} 0$.

Since both $S$ and its denominator $2^{-m}$ are apart from 0, so is its numerator $f(c) - f(c - 2^{-m})$, and therefore we have $f(c - 2^{-m})\ \#\ f(c)$.

For the $x$ in the theorem we can now take $c - 2^{-m}$, if $m > k$, for then $x$ is as close to $c$ as asked. Otherwise, we repeat the proof with a larger value of $m$, setting $m := k + 1$. Increasing $m$ just restricts the original interval to a subinterval with the same endpoint $c$. The relations in step 1 will continue to hold, because they are governed by only a universal quantifier. Also steps 2 and 3 go through as before, and we will obtain $x = c - 2^{-(k+1)}$. That there are these two possibilities to consider for $m$ reflects the fact that a construction for an interval in which all derivatives are close to $f'(c)$ may well yield one that is wider than $2^{-k}$, where $k$ was, after all, chosen arbitrarily.                                                                                                                            $\square$

We will now apply our acquired knowledge to polynomials. For iterated differentiation, we use this notation: $f^{(0)} := f$, $f^{(1)} := f'$, $f^{(n+1)} := (f^{(n)})'$.

**Lemma 5.21** *If $f$ is $n$ times uniformly differentiable on $[a, b]$ with $a < b$, $f(a) < f(b)$, and $|f(x)| + |f'(x)| + \cdots + |f^{(n)}(x)| > 0$ for all $x \in [a, b]$, then there is an $x \in [a, b]$ with $f(x)\ \#\ 0$. If, in addition, $f(a) < 0 < f(b)$, then $f$ has a zero on $[a, b]$.*

**Proof** If $|f(x)| + |f'(x)| + \cdots + |f^{(n)}(x)| > 0$, then there is an $m \leq n$ with $|f^{(m)}(x)| > 0$. We now apply Lemma 5.20: near an $x$ of the interval there is another point $y$ with $|f^{(m-1)}(y)| > 0$. Repeat this argument until we find a $z$ with $|f'(z))| > 0)$. We now simply conclude that within an arbitrarily small distance from the given point $x$ there is a point with function value apart from 0. Based on Theorem 5.8 it now follows that if $f(a) < 0 < f(b)$, $f$ has a zero on $[a, b]$.                                    $\square$

**Theorem 5.22** *If $f$ is a polynomial and $f(0) < 0$, $f(1) > 0$ then $f$ has a zero between 0 and 1.*

**Proof** Let $f(x) = a_n x^n + a_{n-1} x^{n-1} + \cdots + a_0$. The preceding lemma will apply as soon as we have shown $\forall x \in [0, 1]\ (|f(x)| + |f'(x)| + \cdots + |f^{(n)}(x)| > 0)$, which we will do by showing $\inf_{x \in [0,1]}(|f(x)| + |f'(x)| + \cdots + |f^{(n)}(x)| > 0)$. We know that $f(1) = a_0 + a_1 + \cdots + a_n > 0$ and $f(0) = a_0 < 0$. From these we know that $a_n + a_{n-1} + \cdots + a_1 > -a_0 > 0$, whence for some $i$, with $0 < i \leq n$, $a_i\ \#\ 0$. Calculate the $i$-th derivative:

$$f^{(i)}(x) = \frac{n!}{(n - i)!} a_n x^{n-i} + \cdots + i!\, a_i\,.$$

If $i = n$, then $a_n\ \#\ 0$ and $f^{(n)}(x) = n!\, a_n$. In this case,

$$(|f(x)| + |f'(x)| + \cdots + |f^{(n)}(x)| \geq |n!\, a_n|) > 0,$$

whence $\inf_{x \in [0,1]}(|f(x)| + |f'(x)| + \cdots + |f^{(n)}(x)| \geq |n!\, a_n|) > 0)$.

If $i < n$, we appeal to weak trichotomy:

$$\inf |f^{(i)}| < i!|a_i| \vee \inf |f^{(i)}| > \frac{1}{2}i!|a_i| \, .$$

If the second disjunct is true, the same reasoning as in the case $i = n$ yields the desired inequality. If the first disjunct holds, we can find an $x_0$ such that

$$\left| \frac{n!}{(n-i)!}a_n x_0^{n-i} + \cdots + \frac{(i+1)!}{1!}a_{i+1}x_0 + i!\, a_i \right| < i!|a_i| \, .$$

It follows immediately that

$$- i!|a_i| < \frac{n!}{(n-i)!}a_n x_0^{n-i} + \cdots + \frac{(i+1)!}{1!}a_{i+1}x_0 + i!\, a_i < i!|a_i| \, .$$

But $a_i \mathrel{\#} 0$ means $a_i < 0 \vee a_i > 0$. In the first case, we have

$$i!\, a_i < \frac{n!}{(n-i)!}a_n x_0^{n-i} + \cdots + \frac{(i+1)!}{1!}a_{i+1}x_0 + i!\, a_i \, ,$$

whence

$$0 < \frac{n!}{(n-i)!}a_n x_0^{n-i} + \cdots + \frac{(i+1)!}{1!}a_{i+1}x_0 \, ;$$

in the second case, we have

$$\frac{n!}{(n-i)!}a_n x_0^{n-i} + \cdots + \frac{(i+1)!}{1!}a_{i+1}x_0 + i!\, a_i < i!\, a_i \, ,$$

whence

$$\frac{n!}{(n-i)!}a_n x_0^{n-i} + \cdots + \frac{(i+1)!}{1!}a_{i+1}x_0 < 0 \, .$$

Either way we have

$$\frac{n!}{(n-i)!}a_n x_0^{n-i} + \cdots + \frac{(i+1)!}{1!}a_{i+1}x_0 \mathrel{\#} 0 \, .$$

Hence, again $\exists j < i (a_j \mathrel{\#} 0)$ holds. The index has moved forward. By repetition this can be done until $i = 0$ and the condition is met. The previous lemma now applies.

$$\square$$

It is not difficult to see that the theorem holds for any closed interval.

**Corollary 5.23** *The continuum is real closed.*

Definitions of "real closed" vary a little. In traditional algebra it means that every positive $a$ must have a square root and that every odd degree polynomial has a zero. The latter means to us that the leading coefficient, $a_{2n+1}$, is apart from 0, in which case it may be taken to be 1.

## 5.6  Brouwer's Fixed Point Theorem

After earning his doctorate on intuitionistic matters, Brouwer also took up classical topology in such a way that he hoped that his results could eventually be given intuitionistic proofs. His two most famous classical results, the invariance of dimension [19] and the Fixed Point Theorem [20], have, however, eluded such proof, and weak counterexamples show that there is no hope of finding it. But for the classical Fixed Point Theorem, Brouwer found constructive approximations, so to speak: instead of a method that constructs, for suitable $f$, an $x$ such that $f(x) = x$, there is a method that, given an arbitrary distance, constructs a point that $f$ displaces by less than it [25, 26]. In this section we will illustrate all this for the one- and two-dimensional cases, where the discussion of the former serves as preparation for that of the latter. Our proofs are different from Brouwer's. For a comparison of Brouwer's method and the one below (based on Sperner's Lemma), see [78].

### 5.6.1  *The Classical Theorem (One-Dimensional)*

**Theorem 5.24** (Brouwer's classical Fixed Point Theorem) *Let $f$ be a continuous function $[0, 1] \rightarrow [0, 1]$. Then $f$ has a fixed point.*

The proofs in this section and the next two use, for convenience, *barycentric coordinates*, which we pause to introduce here. Let $I$ be a closed interval on the straight line. A point $p$ on that interval will have a Cartesian coordinate given by a real number $x$, with $a \leq x \leq b$, where $a$ and $b$ are the coordinates that the endpoints of $I$ receive. We will assume that $a < b$. In the barycentric system, the position of $p$ is rather given by an  ordered pair of real numbers $\langle x_0, x_1 \rangle$, with $x_0, x_1 \geq 0$ (but not both $x_0 = 0$ and $x_1 = 0$), in a way inspired by the physics of a lever (Fig. 5.14).

Think of the interval $I$ as a straight beam, and place a pivot under it at point $p$. The beam will balance if we place a certain mass of size $x_0$ at the end with coordinate $a$ and one of size $x_1$ at the end with coordinate $b$. Physically, $p$ would be the center of mass of the two attached ones; the term "barycenter" is a synonym for "center of mass" derived from the ancient Greek for "heavy", "βαρὺς". Archimedes' law of the lever states that in this situation, we have the proportion

$$x_0 : x_1 = b - x : x - a . \tag{5.1}$$

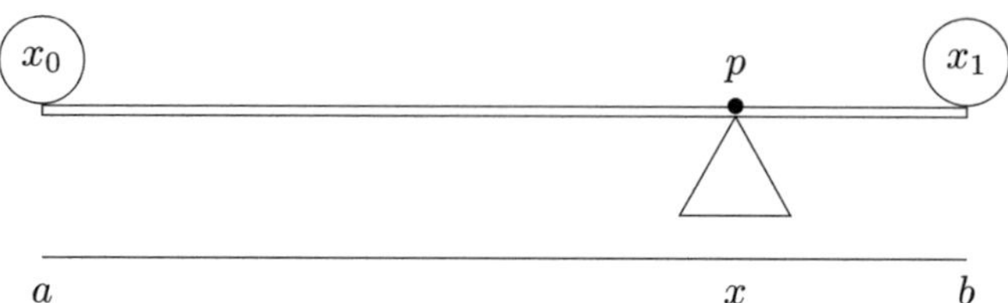

**Fig. 5.14** Physical interpretation of a barycentric coordinate $\langle x_0, x_1 \rangle$ that corresponds to the Cartesian coordinate $x$; either one serves to indicate where on the beam point $p$ lies

Equivalence of two ratios is defined by

$$\langle x_0, x_1 \rangle \sim \langle y_0, y_1 \rangle := \exists r \in \mathbb{R}\ (x_0 = ry_0 \wedge x_1 = ry_1)\,.$$

Similarly for ratios of $n$ numbers. (Such an $r$ is of course unique.) We do not consider $0{:}0$ to be a ratio.

Thus, for given $a$, $b$, and $x$ there will be many different $x_0, x_1$ that satisfy (5.1), but the ratio $x_0 : x_1$ remains the same. Therefore, that ratio characterizes the location of $p$ just as much as its Cartesian coordinate $x$ does. This is verified by

**Theorem 5.25** *There is a $1-1$ mapping between the Cartesian coordinates of points on the interval $I = [a, b]$ and the sets (equivalence classes) of barycentric coordinates $\langle x_0, x_1 \rangle$ whose components stand in a given ratio.*

**Proof** Define the set $B_{\mathbb{R}} := \{\langle r, s \rangle \mid r, s \in \mathbb{R} \wedge \neg(r = 0 \wedge s = 0)\}$, and the functions

$$f : \mathbb{R} \to B_{\mathbb{R}}$$
$$x \mapsto \langle b - x, x - a \rangle$$

and

$$g : B_{\mathbb{R}} \to \mathbb{R}$$
$$\langle x_0, x_1 \rangle \mapsto a + \frac{x_1(b - a)}{x_0 + x_1}$$

Now check that $f$ and $g$ are each other's inverses: $g(f(x)) = x$, and $f(g(\langle x_0, x_1 \rangle)) = \langle x_0, x_1 \rangle$. This induces a bijection between $\mathbb{R}$ and $B_{\mathbb{R}}/_{\sim}$. $\qquad \square$

Barycentric coordinates can be made unique or normalized by requiring that

$$x_0 + x_1 = 1\,. \tag{5.2}$$

This condition can be met by dividing given $x_0$ and $x_1$ by $x_0 + x_1$. (This is permitted, since $x_0$ and $x_1$ are not allowed to be both 0, and both are $\geq 0$.) If $x = a$, then (5.1) entails that $x_1 = 0$, and upon normalization $x_0 = 1$. Similarly, if $x = b$, then $x_0 = 0$ and upon normalization $x_1 = 1$. So to the Cartesian coordinates $a$ and $b$ of the endpoints of $I$ correspond the normalized barycentric coordinates $(1, 0)$ and $(0, 1)$.

To calculate the distance $d(x, y)$ between two points on $I = [a, b]$ given in barycentric coordinates, or, to be precise: the length of the displacement vector, we begin by defining $d((1, 0), (0, 1)) := 1$. Then for arbitrary $x, y \in I$, we first convert them to Cartesian coordinates, calculate the Cartesian distance, and scale the outcome by dividing it by $b - a$.

**Exercise 5.26**  Write down the formula for the Cartesian distance between two points on $I = [a, b]$, given by their barycentric coordinates $x$ and $y$. Check that if we scale it and use normalized barycentric coordinates, we get $d(x, y) = |x_1 - y_1| = |y_0 - x_0|$.

**Exercise 5.27**  Show that the functions $f$ and $g$ in Theorem 5.25 are continuous, and hence the bijection is bi-continuous.

**Remark 5.28**  It follows that if $h \colon \mathbb{R} \to \mathbb{R}$ is continuous, so is $k \colon B_\mathbb{R} \to B_\mathbb{R}$, defined by $k(x) = f(h(g(x)))$. Conversely, if $h \colon B_\mathbb{R} \to B_\mathbb{R}$ is continuous, so is $k \colon \mathbb{R} \to \mathbb{R}$, defined by $k(x) = g(h(f(x)))$. For uniformly continuous functions the analogous statements are true. We conclude that also for proofs involving continuity, it does not matter which of the two coordinate systems we use; that is of course as expected.

The reason why normalized barycentric coordinates are useful when discussing a mapping $f$ from $[a, b]$ into itself is that in that case we know not only (5.2) but also $f(x)_0 + f(x)_1 = 1$, so

$$x_0 + x_1 = f(x)_0 + f(x)_1 . \tag{5.3}$$

It is this invariant that is exploited in the proofs below.

In the theorems in this section, we will take $[a, b] = [0, 1]$, but nothing hinges on that.

Now we are ready to derive the stated theorem.

***Proof*** of Theorem 5.24: Let $f : [0, 1] \to [0, 1]$ be continuous. By Remark 5.28, we may replace this by a function where argument and value are given in normalized barycentric coordinates; we give it the same name $f$. Towards a contradiction, we make the hypothesis that $f$ has no fixed point. Each point of the interval receives one of the labels 0, 1 as follows: the label of point $x$ is $i$ if $f(x)_i < x_i$. Such an $i$ must exist, and be unique:

(1)  by the hypothesis, for no $x$ do we have $f(x)_i = x_i$ for $i = 0, 1$;
(2)  the two remaining possibilities are $f(x)_0 > x_0$, $f(x)_1 < x_1$ and $f(x)_0 < x_0$, $f(x)_1 > x_1$; all other relations would violate the invariant (5.3).

Under these conditions, endpoint $(1, 0)$ receives label 0, and endpoint $(0, 1)$ label 1.

We now consider segmentations $S_n$ in which we iteratively divide the interval that is the domain of $f$ into $2n$ segments: at each iteration, we divide each of the previously obtained segments in two. Claim: for each $n$, one or more of the segments in the segmentation will be full (that is, have the labels 0 and 1). This is shown by induction.

For $n = 1$, we make one division of the domain interval, which by the above is itself full. By the law of the excluded middle, the point at which we divide this interval either received the label 0, or the label 1. In the first case, the segment determined by the dividing point and endpoint $(0, 1)$ is full, in the second case, the segment determined by the dividing point and endpoint $(1, 0)$ is full.

Now assume that each segmentation into $2n$ segments has one or more full segments. Look at one of them upon its division into two, and apply the same reasoning as in the case $n = 1$.

For every $\mathcal{S}_n$, let $z_n$ be any point from a full segment. Since the domain interval is compact, there is a convergent sequence of points $z_n$. Let $z$ be the limit of such a sequence; $z$ is therefore the limit of the endpoints of the corresponding segments, say $a^n, b^n$, assuming for the sake of convenience that those endpoints in that order have been labeled 0, 1. So $z_i$ is the limit of a sequence of points $a_i^n$ with $f(a_0^n) < a_0^n$, and the same holds for the $b^n$. Therefore, because of the continuity of $f$, $f(z)_i \leq z_i$ for $i = 0, 1$. Since, by (5.3), $f(z)_0 + f(z)_1 = z_0 + z_1$, we now have $f(z) = z$, which contradicts our hypothesis that there was no fixed point. So there is a fixed point. $\square$

### 5.6.2  A Weak Counterexample

The above proof uses RAA and PEM, and thus yields no construction method for a fixed point; it is therefore intuitionistically unacceptable. Moreover, it is highly unlikely that a constructive proof can be found, as the Fixed Point Theorem falls victim to a weak counterexample. The argument is really the same as that in the counterexample to the classical Intermediate Value Theorem, Theorem 5.4.

See Fig. 5.15. We use Cartesian coordinates. We here consider the piecewise linear function $f : [0, 1] \rightarrow [0, 1]$ whose graph goes through the points $A = (0, 1/4 + a)$, $B = (1/4, 1/4 + a)$, $C = (3/4, 3/4 + a)$, $D = (1, 3/4 + a)$, where $a$ is one of those numbers, generated in a weak counterexample, which is very close to 0 but for which it is unknown whether $a \geq 0$ or $a \leq 0$ (see Sect. 4.5). The figure of course shows the graph we could draw if we *would* know that $a > 0$.

A fixed point of $f$ occurs where the graph of $f$ intersects the diagonal. Suppose there is a fixed point $u$. Because $1/4 < 3/4$, we also have (by Theorem 4.14) that $1/4 < u$ or $u < 3/4$. In the first case we get $a \geq 0$ and in the second case $a \leq 0$. So the Fixed Point Theorem tells us that we can decide $a \leq 0 \vee a \geq 0$. That is not the case. We conclude that the Fixed Point Theorem is not constructively correct.

### 5.6.3  The Constructive Theorem (One-Dimensional)

However, one can look for a constructive version if by that one means not that a fixed point can be found, but points whose image under $f$ is arbitrarily close to themselves. We have seen something like this before, namely in the approximate Intermediate Value Theorem (Theorem 5.10). There we found that we may not be able to construct a point where a function takes value 0, but also that for every $\epsilon$ there are points with

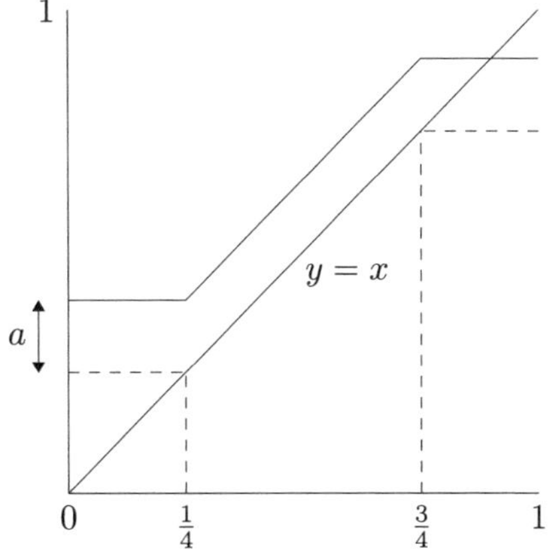

**Fig. 5.15**  If a point of intersection can be determined, we obtain information about $a$

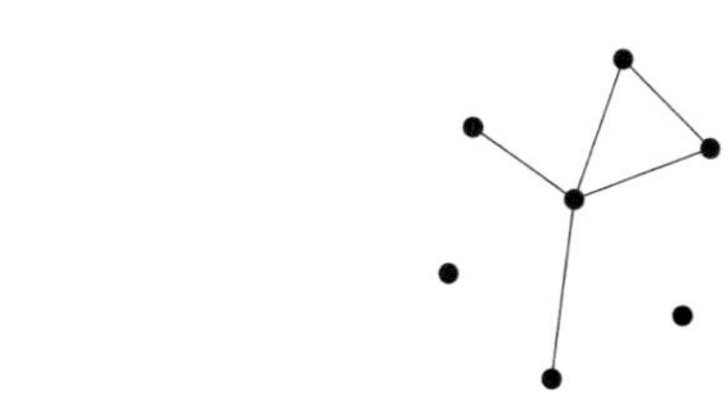

**Fig. 5.16**  A simple graph, with 5 even nodes and 2 odd ones

a function value closer to 0 than $\epsilon$. Indeed, by the remarks after the proofs of that theorem, it is essentially equivalent to the approximate Fixed Point Theorem.[4]

The condition for the approximate Intermediate Value Theorem asks only for continuity. However, we will now prove a version of the approximate Fixed Point Theorem that asks for *uniform* continuity. Our rationale for proving the theorem with a stronger condition, and therefore a weaker theorem, lies not in the result but in the method used to prove it: it introduces the method that will also be used in the more complicated proof for the two-dimensional case in Sect. 5.6.4. For the same purpose, we first present Sperner's Lemma. Its use here admittedly is overkill, but this prepares for its use in the two-dimensional case.

We first prove, as a lemma to the Lemma, a helpful little theorem about a special kind of graphs. A graph in which any two different nodes are connected by at most one edge and in which no node has an edge to itself is called *simple*. We say that a node in it is *even* if it is the endpoint of an even number of edges. A node that is not even is called *odd* (Fig. 5.16).

**Lemma 5.29**  *A simple graph contains an even number of odd nodes.*

***Proof***  Induction to the number of nodes of the graph.

*Basis.* The single node cannot have a loop; so the number of odd nodes is 0.

*Induction step.* Consider a graph $G$ with $n + 1$ nodes. Assume there is an even node $A$, and let $k$ be the number of edges that connect with it. By hypothesis, the

---

[4] As Wim Veldman reminded us.

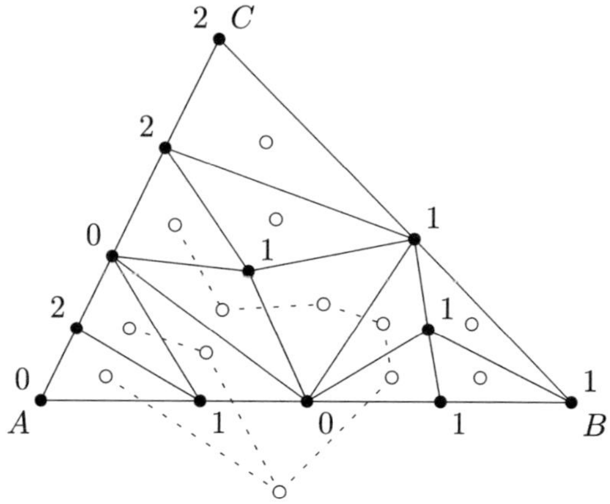

**Fig. 5.17** A Sperner labeling of a triangulation. The inscribed white nodes and dashed edges form a related simple graph that arises in the proof of the Lemma

statement holds for the graph $\mathcal{G}'$ without $A$ and the edges connecting with $A$. Since the other node of each edge connecting with $A$ is either even or odd, $k = p + q$, with $p$ the number of edges going to even nodes, and $q$ the number of edges going to odd nodes. So adding $A$ and its connecting edges to $\mathcal{G}'$ changes $p$ even nodes to odd, and $q$ odd nodes to even. Therefore, $|p - q|$ odd nodes are added or $|p - q|$ such are taken away. As $k = p + q$ is even, so is $p - q$. This proves the fact for $\mathcal{G}$. Consider the case for an odd node $A$ yourself.                                     □

Consider a triangle $ABC$ and triangulate it (Fig. 5.17). That is, divide the triangle into finitely many smaller triangles so that the triangles connect to common sides. All vertices are marked with the number 0, 1, or 2 by a function $f$. We choose $f(A) = 0$, $f(B) = 1$, $f(C) = 2$. The only restriction concerns points on the sides; for $P$ on $AB$, we must have $f(P) = f(A)$ or $f(P) = f(B)$, and analogously for $AC$ and $BC$. We then have a *Sperner labeling*. A triangle from the triangulation is called *full* if the three vertices have the labels 0, 1, 2.

**Lemma 5.30** (Sperner's Lemma) *A  triangulation with a  Sperner labeling has a full triangle.*

***Proof*** For the triangulation, we make a graph whose nodes are the elements of the triangulation, that is, the individual triangles, plus an extra node that lies outside triangle ABC. Two nodes are connected by an edge if the triangles have a 0–1 side in common, and the "outside node" is connected to the nodes in the triangles that have a 0–1 side at the base. This is a simple graph as defined above. We have drawn it into Fig. 5.17, indicating its nodes by a white circle.

Now, only points labeled 0 or 1 can occur on the base $AB$. For the given triangulation there is an odd number of sides with 0–1 endpoints on the base. (Show that yourself.) Therefore, the node outside the triangle is an odd node. By Theorem 5.29, there then also is an odd node inside the large triangle. As from the definition of the edges it immediately follows that from a node within the large triangle 0, 1, or

2 edges start, this means that from an odd node inside exactly one edge goes out, meaning that the small triangle around the node has exactly one side labeled with 0 and 1. Then that small triangle's remaining vertex must have label 2, and hence the small triangle is full.                                                                   □

We now turn to the actual theorem. We will use barycentric coordinates again.

**Theorem 5.31** (Brouwer's one-dimensional constructive (or approximate) Fixed Point Theorem) *Let $f$ be a uniformly continuous function from the unit interval to itself. Then for every $n$ there is an $x$ with $d(x, f(x)) < 2^{-n}$.*

In fact, there is a variety of constructive mathematics in which all functions on a closed interval are uniformly continuous (Brouwer's—see Sect. 6.4), but also one in which there are continuous functions on a closed interval that are not uniformly continuous (Markov's recursive mathematics—see Sect. 6.9.1). So we here make the condition explicit.

***Proof*** By the uniform continuity of $f$ on $[(1, 0), (0, 1)]$, for every $\epsilon > 0$ there is a $\delta > 0$ such that $d(x, y) < \delta \to d(f(x), f(y)) < \epsilon$. It is not a limitation to assume that $\delta < \epsilon$. (Why not?)

We have introduced $\delta$ and $\epsilon$ for convenience; as usual in constructive mathematics, we take rational numbers for these numbers, and here, again as usual, powers of $2^{-1}$: $\delta = 2^{-m}, \epsilon = 2^{-n}$, and by our assumption $m > n$.

Divide $[(1, 0), (0, 1)]$ into segments of length $\delta$. Let $[a, b]$ be such a segment; we divide it into four equal parts (Fig. 5.18).

We then divide the entire interval $[(1, 0), (0, 1)]$ into similar segments of length $2^{-2}\delta$. As in the proof of the classical theorem, we will assign labels to the endpoints of these segments, using Sperner's lemma (thus, a one-dimensional "triangulation"). We want to check for each point whether $f$ moves it to the left or to the right. But we have to keep in mind that a point may also be left in place, because, now reasoning constructively, we cannot give a proof by contradiction from the assumption that there is no such point. So the $<$ should be replaced by $\leq$. However, $f(x) \leq x$ is not decidable, so we must, in characteristic constructive manner, consider small intervals around $x$ instead.

From $x_0 < x_0 + 2^{-4}\delta$ it follows that $x_0 < f(x)_0$ or $f(x)_0 < x_0 + 2^{-4}\delta$. If the first case obtains or both, it follows from $x_0 + x_1 = f(x)_0 + f(x)_1 = 1$ that $f(x)_1 < x_1 + 2^{-4}\delta$, and we assign to $x$ the label 1. If only the second case obtains, the label will be 0. This yields a labeling for all points, labeling $x$ with an $i$ such that

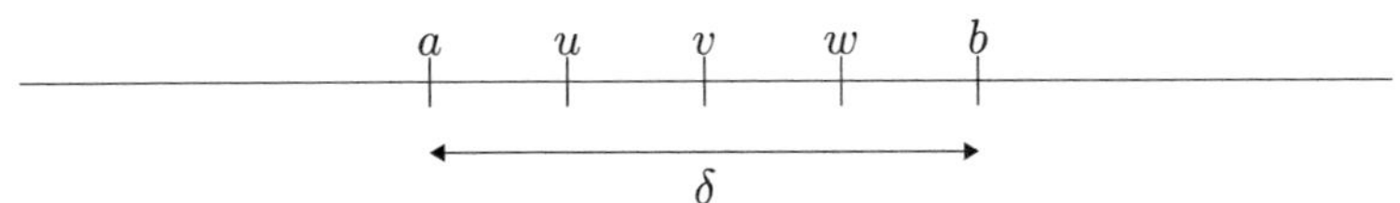


**Fig. 5.18**  Four equal parts

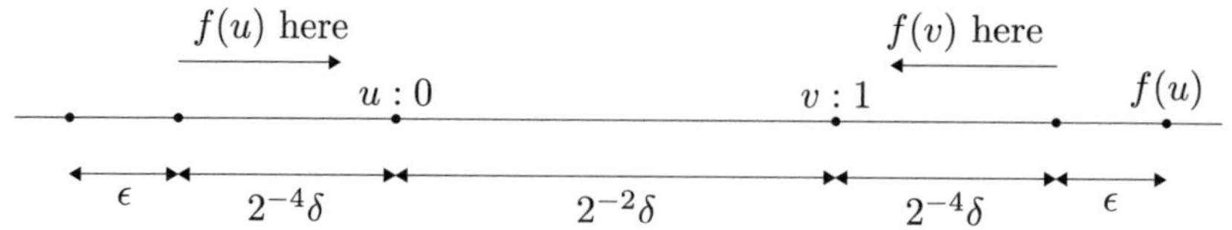


**Fig. 5.19**  A Sperner labeling in the one-dimensional case

$f(x)_i < x_i + 2^{-4}\delta$. The point $(1, 0)$ receives label 0 (why?). By Sperner's lemma, it now is possible to indicate a segment of length $2^{-2}\delta$ with label 0 on the left and label 1 on the right.[5]

Consider such a segment, say $[u, v]$. (Fig. 5.19 is not to scale, as $\delta < \epsilon$.) In determining the labels of $u$ and $v$, we already saw in what parts of the unit interval their respective images under $f$ end up. We now refine this by also taking into account the uniform continuity of $f$, which puts a limit on how far the two images are removed from one another: $d(f(u), f(v)) < \epsilon$. In the figure we have indicated $f(u)$ at its rightmost possibility, or rather just to the right of that. We see that $d(u, f(u)) < 2^{-2}\delta + 2^{-4}\delta + \epsilon < 2^{-1}\delta + \epsilon < 2\epsilon$. This shows that for every $\epsilon$ there is a point that $f$ moves by less than $2\epsilon$.                                                                    $\square$

Observe that the approximate fixed point that we found, $u$, was a point we already constructed on the interval when we segmented it, and we then also calculated $f(u)$ in order to determine a label. So the outcomes of these calculations will have made us realize already then that $u$ is an approximate fixed point. However, that this is *bound* to happen is something that we cannot know before we have proved the general theorem—that is precisely its content.[6]

### 5.6.4  *The Constructive Theorem (Two-Dimensional)*

For the two-dimensional case, we will content ourselves with functions on a triangular region that we will call $\Delta$. This is no loss of generality. While the proof will be more

---

[5] For this case, Sperner's lemma is not really needed, of course. A direct argument for this possibility runs as follows. The segments on [0, 1] are finite in number. The leftmost segment has label 0 on the left. If its label on the right is 1, we have found a segment as asked. Otherwise, it is 0, we move to the segment on its right, and make the same check. Repeating this if needed, it may happen that we get as far as the rightmost segment but one, and find that it has labels 0 and 0. Then the rightmost segment has labels 0 and 1. We could also have argued with induction on the number of times we made a division to create two new intervals anywhere. For 0 times, we have the original interval, which is labeled with 0 and 1. Now assume that we have just divided the whole interval $n$ times and that this yielded a 0–1 interval somewhere. If we do not choose to make the next division in the latter, it remains. If we do choose to divide it, then we obtain, depending on the label of the dividing point, either an interval labeled 0–1 on the left or an interval labeled 0–1 on the right.

[6] This paragraph is in fact a comment on a remark in [78], a review of the earlier Dutch edition of this book.

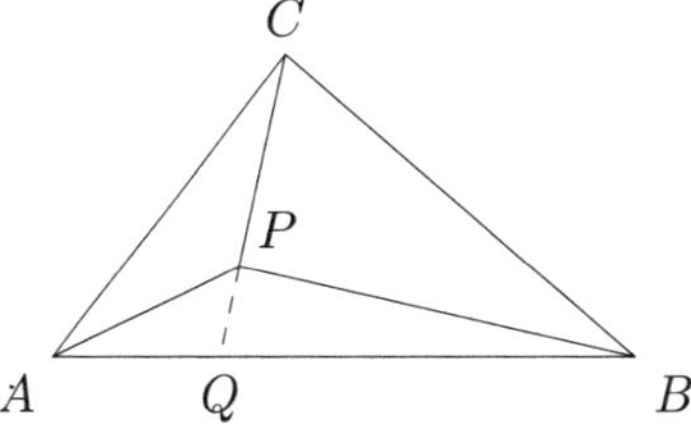

**Fig. 5.20** The triangulation determined by $P$

intricate, the approach is the same as above: divide the domain into smaller pieces ("simplices", singular "simplex"), label their corners according to the displacement of $x$ to $f(x)$, and establish that there is a piece all of whose corners have received a different label. For the labeling we use Sperner's Lemma again.

**Theorem 5.32** (Brouwer's two-dimensional constructive (or approximate) Fixed Point Theorem) *Let $f$ be a uniformly continuous function $\Delta \to \Delta$, where $\Delta$ is a triangular region. Then for every $n$ there is an $x$ such that $d(x, f(x)) < 2^{-n}$.*

In our proof we will again use normalized barycentric coordinates, and assume that $f$ is defined accordingly. For points in two dimensions, the barycentric coordinate system can be motivated analogously to the one-dimensional case, but with three masses. Just as there the coordinates were introduced starting from two lengths ($b - x$ and $x - a$), one dimension higher they can be based on three areas. In a triangle, a point $P$ determines a triangulation (Fig. 5.20).

Let the elements of this triangulation, $\triangle PBC$, $\triangle PCA$, and $\triangle PAB$, have areas $A_{\triangle PBC}$, $A_{\triangle PCA}$, $A_{\triangle PAB}$. If $P$ lies on a side, say $AB$, then $\triangle PAB$ is degenerate, having area 0; if $P$ is a vertex, two of these triangles have area 0; otherwise all three areas are non-zero.

Projecting $P$ onto $AB$ along the line $CP$ yields the point $Q$. Choose $x_0$ and $x_1$ such that $x_1 : x_0 = QA : QB$. Note that $(x_0, x_1)$ would be the barycentric coordinate of $Q$ if we only considered $AB$. Using the fact that the area of a triangle is half the product of the base and height, we have $A_{\triangle QAC} : A_{\triangle QBC} = x_1 : x_0 = A_{\triangle PAQ} : A_{\triangle PBQ}$. It follows that $A_{\triangle PAC} : A_{\triangle PBC} = x_1 : x_0$.

By applying the same reasoning to projections of $P$ onto $AC$ and onto $BC$, we obtain

$$A_{\triangle PBC} : A_{\triangle PAC} : A_{\triangle PAB} = x_0 : x_1 : x_2 .$$

If we furthermore normalize areas so that $A_{\triangle ABC} = 1$, and set $x_i' = x_i/(x_0 + x_1 + x_2$, we get

$$A_{\triangle PBC} + A_{\triangle PAC} + A_{\triangle PAB} = x_0' + x_1' + x_2' = 1 .$$

It is now easily proved that to distinct points belong distinct triples $(x_0', x_1', x_2')$: again using the formula for the area of a triangle, we see that any point $R$ yielding the same areas $A_{\triangle PBC}$, $A_{\triangle PCA}$, $A_{\triangle PAB}$ must lie on the lines passing through $P$ parallel to the sides. Thus, $R$ coincides with $P$, the unique point of intersection of these three lines. Hence, $(x_0', x_1', x_2')$ serves to define the normalized barycentric coordinate of $P$.

**Fig. 5.21** Cartesian and normalized barycentric coordinates

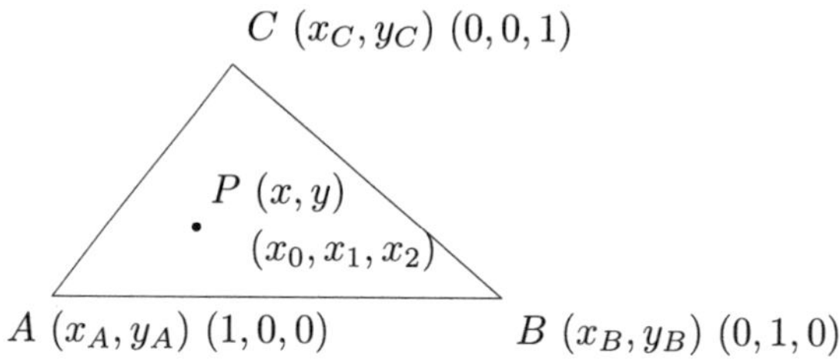



As in the one-dimensional case, there are continuous functions to convert from Cartesian to  barycentric coordinates and vice versa; we here simply state them.

See Fig. 5.21. Define the set $B_{\mathbb{R}^2} := \{\langle r, s, t\rangle \mid r, s, t \in \mathbb{R} \wedge \neg(r = 0 \wedge s = 0 \wedge t = 0)\}$, and the functions

$$f : \mathbb{R}^2 \to B_{\mathbb{R}^2}$$
$$(x, y) \mapsto \langle x_0, x_1, x_2\rangle$$

with

$$x_0 = \frac{(y_B - y_C)(x - x_C) + (x_C - x_B)(y - y_C)}{(y_B - y_C)(x_A - x_C) + (x_C - x_B)(y_A - y_C)},$$

$$x_1 = \frac{(y_C - y_A)(x - x_C) + (x_A - x_C)(y - y_C)}{(y_B - y_C)(x_A - x_C) + (x_C - x_B)(y_A - y_C)},$$

$$x_2 = 1 - x_0 - x_1,$$

and

$$g : B_{\mathbb{R}^2} \to \mathbb{R}^2$$
$$\langle x_0, x_1, x_2\rangle \mapsto (x, y)$$

with

$$x = x_0 x_A + x_1 x_B + x_2 x_C,$$
$$y = x_0 y_A + x_1 y_B + x_2 y_C.$$

Similarly to the one-dimensional case, the functions $f$ and $g$ are each other's inverses, and this induces a bi-continuous bijection between $\mathbb{R}^2$ and $B_{\mathbb{R}^2}/_{\sim}$.

The calculation of distances proceeds similarly: we convert the coordinates to Cartesian ones, determine the Cartesian distance, and scale back.

***Proof of Theorem*** 5.32 The argument is a bit more involved than what we have seen in the one-dimensional case, yet wholly parallel to it.[7] In fact, this was our reason for presenting the one-dimensional case the way we did.

Because any two triangles can neatly and continuously be transformed into one another, we will not attempt to reason about an "arbitrary triangle", as we are now free to choose a triangle that is convenient for the proof. We take an equilateral triangle

---

[7] A first version of this proof was presented in [98].

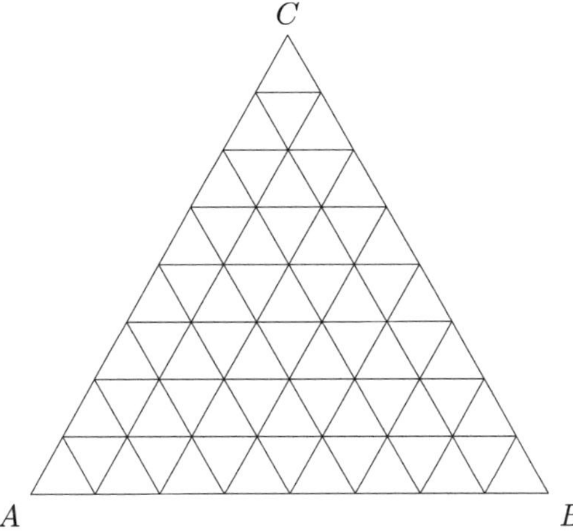

**Fig. 5.22** The triangulation $\mathcal{T}_3$, which has $2^3$ small triangles along each side

$\Delta$ with sides of length 1; the reason why that is convenient will become clear at the end of the proof, when the time has come actually to express a bound on $|f(x) - x|$. The vertices $A, B, C$ are at $(1, 0, 0)$, $(0, 1, 0)$, and $(0, 0, 1)$.

The standard formulation for uniform continuity is:

$$\forall \epsilon \exists \delta \forall x \forall y (d(x, y) < \delta \rightarrow d(f(x), f(y)) < \epsilon)$$

For our proof we take $\epsilon = 2^{-n}$ and $\delta = 2^{-k} + 2^{-k-2}$. It goes without saying that we can do this in such a way that $\delta < \epsilon$. We introduce another abbreviation: $\tau = 2^{-k-3}$, so $\delta = 2^{-k} + 2\tau$.

The triangulation $\mathcal{T}_k$ we use is obtained by dividing the sides into $2^k$ equal parts and drawing lines through those points parallel to the sides. This divides $\Delta$ into triangles with sides of length $2^{-k}$ (Fig. 5.22).

Towards an application of Sperner's Lemma, we want to construct labels for the vertices of the small triangles. As in the proof of the approximate Fixed Point Theorem for dimension 1, the labels cannot be chosen in exactly the same way as in the classical case, because we cannot assume (for a proof from contradiction) that there is no fixed point, and because the ordering is, after all, undecidable.

From $x_0 < x_0 + \tau$ it follows that $x_0 < f(x)_0$ or $f(x)_0 < x_0 + \tau$. If $x_0 < f(x)_0$, it follows from $x_0 + x_1 + x_2 = f(x)_0 + f(x)_1 + f(x)_2 = 1$ that $x_1 + x_2 > f(x)_1 + f(x)_2$, and so $x_1 > f(x)_1$ or $x_2 > f(x)_2$. (Here we applied Theorem 4.41(10), as we will do again in this proof.) Hence, for every $x$: $f(x)_0 < x_0 + \tau$ or $f(x)_1 < x_1 + \tau$ or $f(x)_2 < x_2 + \tau$.

So we can label $x$ with an $i$ such that $f(x)_i < x_i + \tau$. We make sure to label $A$ with 0, $B$ with 1, and $C$ with 2.

Now we check the condition of Sperner's Lemma on the nodes on the sides of $\Delta$. Take an $x$ that is on side $AB$, so $x_2 = 0$. The above procedure yields a label $i$ with $f(x)_i < x_i + \tau$. We show: if the procedure yields 2, either 0 or 1 would be correct too. Suppose we have a label 2 by virtue of $f(x)_2 < x_2 + \tau$. Since $x_2 = 0$, we get $f(x)_2 < \tau$. Together with $x_0 + x_1 + x_2 = f(x)_0 + f(x)_1 + f(x)_2$ this gives $x_0 + x_1 < f(x)_0 + f(x)_1 + \tau$, from which it follows that

$$x_0 < f(x)_0 \vee x_1 < f(x)_1 + \tau.$$

In the first case we have, as above, $f(x)_1 < x_1$. In the second case, we aim to show that $f(x)_0 < x_0 + \tau$. We begin with

$$x_0 + x_1 + \tau = f(x)_0 + f(x)_1 + f(x)_2 + \tau,$$

from which, since $x_1 < f(x)_1 + \tau$, it follows that $x_0 + \tau > f(x)_0 + f(x)_2$. Setting $u = \tau - f(x)_2$ for brevity, we obtain $x_0 > f(x)_0 + f(x)_2 - \tau = f(x)_0 - u$, and, noting that $\tau > u > 0$, now $f(x)_0 < x_0 + u < x_0 + \tau$. This shows that we can always label the nodes on $AB$ with a 0 or a 1. The arguments for sides $AC$ and $BC$ are analogous.

We apply Sperner's Lemma, and conclude: *every $\mathcal{T}_k$ has a full element.* In Fig. 5.23 we have indicated one, the triangle $uvw$. There are $3 \times 2 \times 1 = 6$ possibilities for labeling each of its vertices differently. Here we see the case $u : 0$, $v : 1$, $w : 2$, which we will treat first. The labels are consequences of $f(x)_i < x_i + \tau$ $(i = 0, 1, 2)$.

We will show that a vertex of a full triangle can only be moved slightly.

We first consider for each of $u$, $v$, $w$ separately in what part of $\Delta$ its image point ends up. The points with barycentric 0-coordinate smaller than $u_0$ are the points that lie to the right of the line that goes through $u$ and is parallel to the opposite side of the large triangle.[8] We have indicated that line with dashes. The line $\ell_u$ is that dashed line shifted such that the points lying to $\ell_u$'s right are those with 0-coordinate smaller than $u_0 + \tau$. The lines $\ell_v$ and $\ell_w$ are obtained in the same way. Now $f(u)$ is in the part of $\Delta$ to the right of $\ell_u$, $f(v)$ is in the part to the left of $\ell_v$, $f(w)$ is in the part below $\ell_w$.

Next, we refine this by taking into account also the uniform continuity of $f$. Since $d(u, v), d(u, w) < \delta$, we have $d(f(u), f(v)) < \epsilon$ and $d(f(u), f(w)) < \epsilon$, so $f(u)$ cannot go arbitrarily far to the right or arbitrarily far up. Since $f(v)$ is bounded on the right by $\ell_v$, $f(u)$ is bounded on the right by $\ell^v$, which is at distance $\epsilon$ from $\ell_v$. Likewise, if $f(w)$ is below the line $\ell_w$, then $f(u)$ must remain below the line $\ell^w$. So $f(u)$ has to remain inside the triangle formed by $\ell_u, \ell^v, \ell^w$. The point at the intersection of $\ell^w$ and $\ell^v$ is therefore certainly farther apart from $u$ than $f(u)$ can be. In Fig. 5.23, we have put $f(u)$ at this point itself, but that is no problem, as what we are looking for is only a bound on its distance from $u$.

The straight line from $u$ to $f(u)$ as drawn in the figure exactly traverses the longer diagonals of 3 adjacent parallelograms, each composed of two equilateral triangles.

---

[8] The consideration here is the same as the one that led to our labeling. If for points $u$ and $x$ we have $x_0 < u_0$, then $x_1 + x_2 > u_1 + u_2$, because of the invariant $u_0 + u_1 + u_2 = x_0 + x_1 + x_2 = 1$. This means that either $x_1 > u_1$, or $x_2 > u_2$, or both, so to get from $u$ to $x$ one has to move towards the top vertex $C$ of the large triangle, or towards its bottom right vertex $B$, or both.

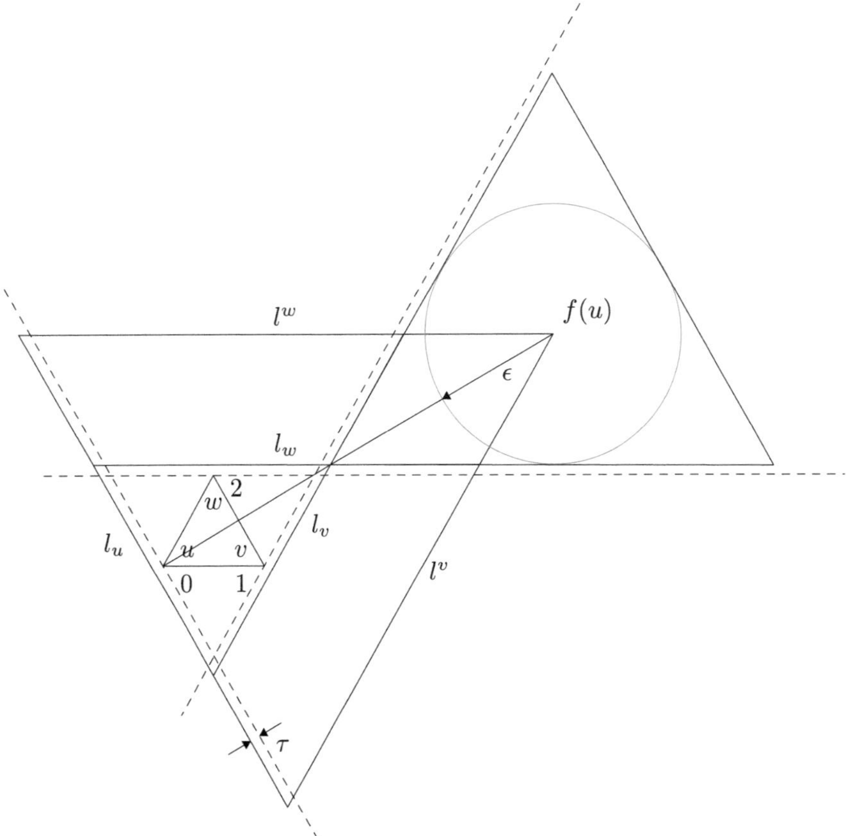



**Fig. 5.23** An example of a full small triangle $uvw$ and $f(u)$, with $f(u)$ drawn just beyond the greatest distance from $u$ allowed by the various conditions

The first has a long diagonal of length smaller than $2^{-k} + 2^{-k}$. The second, small parallelogram has been enlarged in Fig. 5.24. It has a long diagonal of length $2\tau$. The third parallelogram has a long diagonal of length $2\epsilon$: $f(u)$ is the center of the inscribed circle with radius $\epsilon$ in an equilateral triangle, so the length of the bisectors in that triangle is $3\epsilon$, and the length of the diagonal we are looking at is $2/3$ of that. We should like to determine a bound in terms of $\epsilon$ only. To obtain this, we first use the definitions of $\delta$ and $\tau$ to bound $d(u, f(u))$ by $2\delta + 2\epsilon$. With $\delta < \epsilon$ we now obtain $d(u, f(u)) < 4\epsilon$.

Of the 6 possible labelings of the full triangle $uvw$, we have here treated the so to speak worst case. In the others, a similar analysis of the different allowed regions for the points $f(u)$, $f(v)$, $f(w)$ yields a bound on the distance from $u$ to $f(u)$ that is not greater than the one found above, $4\epsilon$. So for given $\epsilon$, we can always find an "almost fixed point" of $f$.                                                                                    $\square$

**Fig. 5.24** $PQRS$ is the
small parallelogram at the
intersection of $l_v$ and $l_w$ in
Fig. 5.23

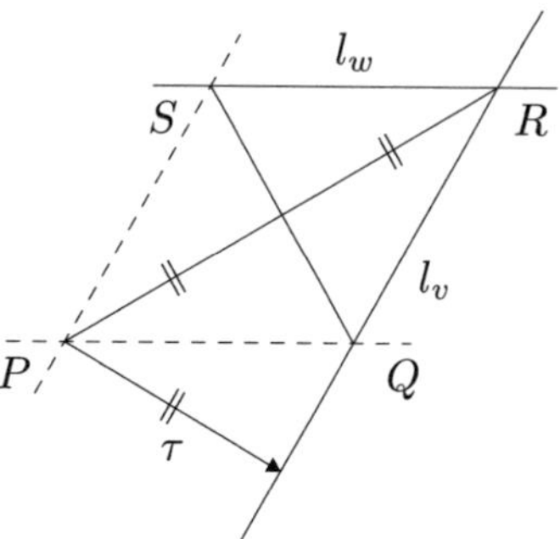


An obvious question now is: can we extend the hypothesis of the theorem so that a
true fixed point can be constructed? That would be similar to the way the Intermediate
Value Theorem can be treated, see [83, pp. 292–294]. So far, no solution has been
found.

# Chapter 6
# Going Forth

**Abstract** We discuss specifically intuitionistic ideas, based on Brouwer's "second act": choice sequences and species. A particular kind of species, spreads, is used to model the non-denumerable continuum. The Continuity Principle, which essentially states that a total function on a spread depends only on finite initial segments of choice sequences, leads us to the Fan Theorem (a form of compactness) and the Uniform Continuity Theorem. We also discuss the Brouwer-Kripke Schema, which models the Creating Subject and has surprising mathematical consequences. Finally, we compare Brouwer's approach with the recursive mathematics of Markov and with the pragmatic but watered-down constructivism of Bishop.

We have now shown a small part of constructive analysis; it lies in the "neutral" part of constructive mathematics, in the sense that what we have done thus far will be recognized immediately by everyone as effective. Constructive mathematics nowadays extends much, much further, and in different directions; see the References and further reading. In this chapter we will turn to various specifically intuitionistic ideas. At the end there will be a brief comparison with the two other major approaches to constructive mathematics: the algorithmic one of Markov, and the pragmatic one of Bishop, which proposes to stay within the neutral part.

## 6.1 The Basis of Brouwer's Intuitionistic Mathematics: The Two Acts

To Brouwer, mathematics is a human construction, for which the basic building material comes from our awareness of the passage of time. It is worth seeing how he describes this in his own words. Here he is in "Points and spaces" from 1954:

*The first act of intuitionism* completely separates mathematics from mathematical language, in particular from the phenomena of language which are described by theoretical logic. It recognizes that mathematics is a languageless activity of the mind having its origin in the basic phenomenon of the perception of a *move of time*, which is the falling apart of a life moment into two distinct things, one of which gives way to the other, but is retained by memory. If the two-ity thus born is divested of all quality, there remains the common



D. van Dalen et al., *Intuitionistic Analysis*, Springer Undergraduate Mathematics Series,
https://doi.org/10.1007/978-3-032-16491-9_6

substratum of all two-ities, the mental creation of the *empty two-ity*. This empty two-ity and the two unities of which it is composed, constitute the *basic mathematical systems*. And the basic operation of mathematical construction is the *mental creation of the two-ity of two mathematical systems previously acquired*, and the consideration of this two-ity as a new mathematical system. [27, p. 2, original emphasis]

Mathematical thought has to begin with two-ity, as unity would not be enough to get the process going. As Brouwer remarked in his dissertation:

The first act of construction has *two* discrete things thought together [...]; F. Meyer [...] says that *one* thing is sufficient, because the circumstance that I think of it can be added as a second thing; this is false, for exactly this *adding* (i.e., setting it while the former is retained) *presupposes the intuition of two-ity*; only afterwards this simplest mathematical system is projected on the first thing and the *ego which thinks the thing*. [28, p. 97n1, original emphasis][1]

This also means that, in the order of construction, unity can only be arrived at after two-ity:

The inner experience (roughly sketched):
twoity;
twoity stored and preserved aseptically by memory;
twoity giving rise to the conception of invariable unity;
twoity and unity giving rise to the conception of unity plus unity;
threeity as twoity plus unity, and the sequence of natural numbers; [...] [30, p. 90]

When a twoity is given originally, it has an earlier part and a later part that are given together but distinctly. What connects them is the continuum of intuitive time, whose passing is an intrinsic part of our awareness. Unity arises when we consider a part separately. That is an act of abstraction: however that works exactly, it involves no longer using (disregarding) some of what was given to us in intuition.

After the "first act of intuitionism" comes a second one, which begins as follows:

*The second act of intuitionism* recognizes the possibility of generating new mathematical entities:
First, in the form of infinitely proceeding sequences whose terms *are chosen more or less freely from mathematical entities previously acquired*; in such a way that the freedom existing perhaps at the first choice may be irrevocably subjected, again and again, to progressive restrictions at subsequent choices, while all these restricting interventions, as well as the choices themselves, may, at any stage, be made dependent on possible future mathematical experiences of the creating subject; [...] [27, p. 2, original emphasis]

The "infinitely proceeding sequences" are also known as "choice sequences". In Brouwer's published writings, the development of their theory begins in a long article, [21]. The idea is simply this. In intuitionism, the objects of mathematics exist only in so far as the subject has itself constructed them. In particular, an infinite sequence exists only in so far as the subject constructs its elements one after the other. When it sets out to construct the $n$-th element in a sequence, it may of course turn to a law

---

[1] For the original Dutch: [17, p. 179n1].

that supplies a value. But the subject can just as well obtain a value by making a free choice. Either way the construction of the sequence can proceed. Choice sequences are not limited to sequences where every element is chosen wholly freely; one may also impose restrictions on choices.

Brouwer's term "infinitely proceeding sequences" is neutral between sequences that involve free choice and sequences that do not. But in the literature it has become customary to consider lawlike sequences a limiting case of choice sequences: they are choice sequences that have been restricted so narrowly that the number of available choices for each element is only 1. The other extreme is that of the lawless sequences, which have the property that no restrictions will ever be imposed on the choices. Anything in between is also possible. Whether, in the construction of any choice sequence, restrictions will come into play, is itself subject to the Creating Subject's free will.

On choice sequences one can operate just as with other sequences—termwise addition, multiplication, etc.

In analysis, they can be used to create Cauchy sequences and, so long as the Cauchy condition is satisfied, the choice of elements is still completely free.

One will also need to be able to speak of all kinds of *collections* of choice sequences; in topology of collections that are open, closed, perfect, etc.; in algebra we use groups, subgroups, ideals, etc. The possibility of collecting choice sequences or other mathematical objects is described in the next part of the "second act":

> Secondly, in the form of mathematical *species*, i.e., *properties supposable for mathematical entities previously acquired*, and satisfying the condition that, if they hold for a certain mathematical entity, they also hold for all mathematical entities which have been defined to be *equal* to it, equality having to be symmetric, reflexive, and transitive, and the empty two-ity being forbidden to be equalized to an empty unity. Mathematical entities for which the property in question holds, are called *elements* of the corresponding species. [27, p. 2, original emphasis]

We will reflect for a moment on Brouwer's changing (widening) views on constructive collections.

In its use in everyday mathematics, the notion of set is a fairly neutral one. There are two ways of specifying a set. Either one specifies collections by some property of its elements. Think of the "locus", e.g., all points that are equidistant from two fixed points, or take the set of the zeros of a function. Or one presents a collection by showing its elements—literally, for example $\{1, 2, 3\}$ or by indicating how to generate them, as in "Go through the natural numbers, and for each $n$, put $2n$ in the set". While such examples are clear enough, others create notorious problems. Just think of the set of all sets, which can and cannot be extended; or of the Russell set, the set of all sets that do not contain themselves as elements, and therefore contains itself if and only if it does not contain itself. In practice, it is rare to run into such problems. However, their existence shows that fundamental concepts are not fully understood, which is not only unsatisfactory theoretically, but also means that bad surprises are not yet excluded.

Ever since Cantor introduced sets as objects of pure mathematics in their own right, existing independently of other kinds of mathematical objects, sets have increasingly come to function as the fundamental objects of all mathematics. Constructivists, too, sought to incorporate Cantor's legacy into their practice. Brouwer thought about the question what sort of collections could be created within his intuitionistic program, and what role they could play, from his 1907 dissertation onward. At the time, his view was that all collections have to be constructed step-by-step, generating their elements out of prior objects, starting from the natural numbers, themselves given as a sequence in the basic intuition ("the first act of intuitionism", above). Hence, sets could be created as finite or countable collections.

This method may seem a bit restrictive, because with sets of the form $\{a_i \mid i \in \mathbb{N}\}$ the possibilities are soon exhausted. It wasn't that bad, actually; the method of countable generation also allows a collection, once recognized as an object that can be generated, to be itself an element in further generation. Consider the set of natural numbers with their natural order, $\omega$. We can now generate a new ordered set by putting two copies of $\omega$ one after the other, obtaining $\omega + \omega$. Within each copy the ordering remains the original one, and we complete the ordering by defining every element of the second copy to be greater than every element in the first. This process can be iterated further, and, in combination with the iteration of adding 1, yields the so-called well-ordered collections of natural numbers. Each of these collections is still countable, but they can have a complicated ordering structure compared to that of $\omega$. This is the beginning of the intuitionistic theory of ordinals. However, no constructive sense can be made of uncountable collections yet; that was the next development.

Brouwer came to realize that in his foundations, too, objects could be collected not only by generating them, but also according to a property. (The former then becomes a special case, as there is the property of having been generated in a certain way.) These are the *species* in the quotation above. They do not, as in classical mathematics, partition an independently existing universe, the objects with the property on the one side and those without it on the other. Rather, species are defined relative to the objects one already has "acquired", as Brouwer writes: the objects one already has constructed, or at least has devised a construction method for. A species is therefore analogous to a set formed by Zermelo's axiom of separation ("Aussonderung") in classical set theory. Even if mathematically some species are highly capricious and uncontrollable things (see the beginning of Sect. 4.8), it is only natural to classify the things you have according to some property.

This dependence of species on a prior domain also requires that, whenever the universe is extended with new objects that would not have been generated by a construction method already in possession, species be defined anew, now relative to the larger universe. Clearly, this way there can be nothing like "the set of all sets" or the Russell set. One says that Brouwer's species are *predicative*.

What makes the notion of species so fruitful in intuitionism is that a property need not consist in some relation in which a constructed object may stand to others or to itself, but may also be an aspect of the *construction process* in which it came about, its mode of generation. For example, there can be no algorithm or law that enumerates all choice sequences of natural numbers (see the end of Sect. 6.3), but we can form the species of sequences that have been constructed by successively choosing natural numbers. Similarly, we cannot enumerate the Cauchy sequences, but we can form the species of sequences that have been constructed so as to satisfy the Cauchy condition. Of decisive mathematical importance is the fact that those species, while infinite, are not countable: that is, we can construct all kinds of infinite lists of elements of them, but we cannot construct one list containing them all. In the next section, we will see how Brouwer made use of this fact to give a constructive theory of the uncountable real numbers, using a special kind of species called *spreads*.

## 6.2  Spreads and Fans

It is true that the basic intuition provides the (temporal) continuum as a whole, but, as Brouwer noted, the continuum could not be seen as a whole that is composed out of discrete elements. As he put it in 1907,

> Since continuity and discreteness occur as inseparable complements, both having equal rights and being equally clear, it is impossible to avoid one of them as a primitive entity, trying to construe it from the other one, the latter being put forward as self-sufficient; in fact it is impossible to consider it as self-sufficient. Having recognised that the intuition of continuity, of "fluidity", is as primitive as that of several things conceived as forming together a unit, the latter being at the basis of every mathematical construction, we are able to state properties of the continuum as a "matrix of points to be thought of as a whole". [28, p. 17]

But at the time he did not manage to give a more specific description of the structure of the continuum in mathematical terms: all he had was the basic intuition, a collection of real numbers given by laws, and the realization that the latter did not fully describe the former. Only in the second phase of his intuitionistic program did Brouwer find the concept that enabled him to make constructive sense even of uncountable species: the concept of spread. With spreads, Brouwer managed to turn choice sequences from something that outsiders considered a curiosity (and that is worded kindly) into highly useful objects.

A spread is, roughly speaking, a species of choice sequences making up a tree. Here are the relevant definitions:

**Definition 6.1**  A *tree* is a partially ordered collection such that

(1)  there is a largest element, the *top*,
(2)  each element is reachable via finitely many direct *successor steps* from the top,
(3)  incomparable elements have no common successors.

Note:

(1)  Trees "grow" down here. What is also called the *root* is here called the *top*.
(2)  A tree has its own natural topology. The basic open species of nodes are the species of nodes less than or equal to a given node, that is, are initial segments of it (not necessarily proper).

**Definition 6.2**  A *spread tree* is a decidable species $S$ of finite sequences (including the empty sequence, ( )) of natural numbers that is closed under predecessors, and in which each node has at least one successor. The natural partial ordering between finite sequences is indicated by $s \prec t$, that is, $t$ is an extension of $s$. More formally:

(1)  $( ) \in S$,
(2)  $\forall s (s \in S \vee s \notin S)$,
(3)  $\forall s \forall t (s \in S \wedge t \prec s \rightarrow t \in S)$,
(4)  $\forall s {\in} S \; \exists t {\in} S \; (s \prec t)$,
(5)  A *choice sequence* (or, in short, *sequence*) over $S$ is an infinite path through the spread tree,
(6)  A spread tree $S$ determines a *spread* consisting of all choice sequences $\alpha$ over (or through) $S$.

The species $S$ is called the *underlying tree* of the spread, and the infinite branches are called the *elements* of the spread.

A spread is called a *fan* if each node of the underlying tree has only finitely many direct successors.

Note:

(1)  A spread tree is a special tree; at least one infinite branch passes through each node.
(2)  A spread tree also has a natural topology: the topological space now consists of the choice sequences, and the base open species consist of all sequences through a given node. This topology is relevant to us.

Important examples are the "universal spread", which admits all choice sequences of natural numbers, and the "binary fan", i.e., the collection of all 0–1 choice sequences, See Figs 6.1 and 6.2. The former can be seen as the collection of all functions from $\mathbb{N}$ to $\mathbb{N}$, and the tree of all finite sequences that carries it is denoted by $\mathbb{N}^{<\mathbb{N}}$. In topology, the space of the choice sequences through that tree, that is, $\mathbb{N}^{\mathbb{N}}$, is called *Baire space* (i.e., the space of all infinite sequences of natural numbers, or of all functions from $\mathbb{N}$ to $\mathbb{N}$). For the binary fan the notation $2^{\mathbb{N}}$ is used, and the underlying binary tree is denoted by $2^{<\mathbb{N}}$. The topological space associated with the binary fan is called *Cantor space*. The paths through these two spreads correspond to real numbers and are not denumerable: this can be shown by Cantor's famous diagonal argument, or with Brouwer's Continuity Principle, which we will look at next.

## 6.3   The Continuity Principle

Brouwer's first major insight here was the Continuity Principle, which comes in different versions. Its simplest case is about functions defined on the universal spread and outputting natural numbers. Such a function receives as input a choice sequence.

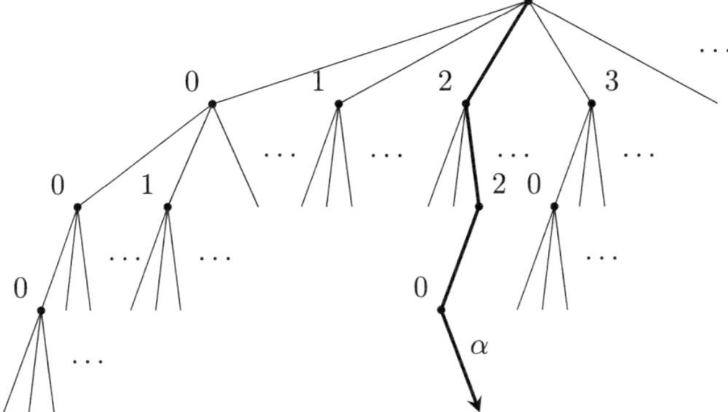

**Fig. 6.1**  The universal spread, and a choice sequence $\alpha = 2, 2, 0, \ldots$ through it

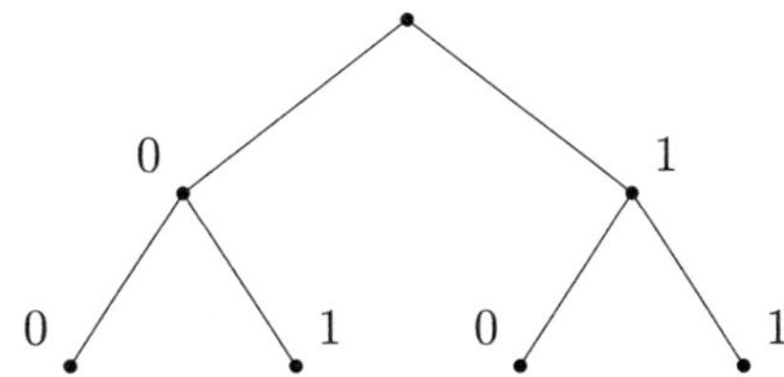

**Fig. 6.2**  Top levels of the binary fan

As a choice sequence is a special kind of function from $\mathbb{N}$ to $\mathbb{N}$, a function on the universal spread is a function whose arguments are themselves functions. In the tradition of the foundations of mathematics such a function is called a *functional* of type $\mathbb{N}^{\mathbb{N}} \to \mathbb{N}$.[2]

To see how such a function works, we present it as a two-person game, with people $A$ and $B$. $A$ is the one who generates the choice sequence $\alpha$, and $B$ is the one who calculates the function value $F(\alpha)$. In the left column of a table, $A$ successively enters his choices. After every choice $B$ continues his calculation, using the new information. When $B$ has received enough information to give the output in the right column, the game is over. As long as $B$ is not yet able to give the answer, $B$ asks $A$ for the next number in the sequence. Now it is given beforehand that the output is a natural number, and a natural number has a finite construction—perhaps with detours, but it has to stop at some point. $B$ has an algorithm in mind to follow, and that leads to requests for new information when needed.

---

[2] Functionals can in turn serve as arguments for functionals.

Let us take a simple example where $F(\alpha)$ is the sum of the first $\alpha(0)$ terms.

| A | B |
|---|---|
| 6 | ? |
| 2 | ? |
| 23 | ? |
| 0 | ? |
| 4 | ? |
| 111 | $F(\alpha) = 146$ |

In this case it is clear that for every $\alpha$ the calculation proceeds in finitely many steps and that only values from the initial segment of $\alpha$ generated at the time of the successful calculation are used. Brouwer's analysis of "calculation of $F(\alpha)$" applies to the general case: *the value of $F$ at input $\alpha$ is determined by a finite initial segment of $\alpha$*. An important condition for this to be so is that $A$ only provides numerical information.[3] Now the conclusion:

**Theorem 6.3** (Continuity Principle) *Every $F : \mathbb{N}^{\mathbb{N}} \to \mathbb{N}$ is continuous.*

We restate the principle using some new notation. We represent the initial segment $\alpha(0), \alpha(1), \ldots, \alpha(n-1)$ of $\alpha$ as $\overline{\alpha}n$. In particular, $\overline{\alpha}0$ is the empty sequence. As a rule, logical literature encodes finite sequences as natural numbers. That is why in logic one simply quantifies over natural numbers when dealing with finite sequences. That is not so important to us now, but it is good to keep in mind that finite sequences behave in the same way as natural numbers in terms of nature and complexity. The Continuity principle can now be formulated as:

$$\forall \alpha \exists x \, F(\alpha) = x \to \forall \alpha \exists m \exists x \forall \beta [(\overline{\beta}m = \overline{\alpha}m \to F(\beta) = x)]$$

where $\alpha$ and $\beta$ range over choice sequences, and $m$ and $x$ over natural numbers. For relations more generally:

$$\forall \alpha \exists x \, A(\alpha, x) \to \forall \alpha \exists m \exists x \forall \beta [\overline{\beta}m = \overline{\alpha}m \to A(\beta, x)]$$

The geometric representation of the universal spread in Fig. 6.1 helps us see what this continuity means. Choice sequences are here the infinite paths through the tree. The Continuity Principle says that a function $F$ that is defined on all these choice sequences determines, for each of its arguments $\alpha$, a node by which $F(\alpha)$ can be constructed. As a consequence, all other paths that go through that node—choice sequences that share that initial segment—will have the same value when $F$ is applied to them. See Fig. 6.3.

Topologically speaking: $F$ is continuous in the natural topology of Baire space. The basic open species are just the collections of functions that pass through an initial segment $k_0, k_1, \ldots, k_{n-1}$. So there is a basic open species of the Baire space,

---

[3] For a discussion and justification of that condition, see [92].

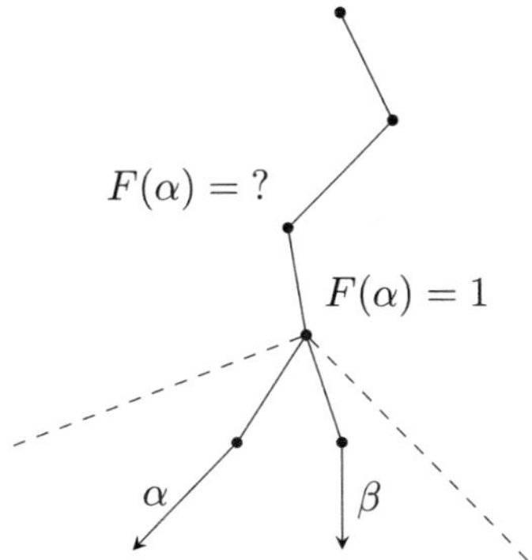

**Fig. 6.3**  The Continuity Principle, by which one concludes that also $F(\beta) = 1$



consisting of all choice sequences $\alpha$ with a certain initial segment $k_0, k_1, \ldots, k_{n-1}$, all elements of which are mapped by $F$ to the same output $F(\alpha)$.

**Exercise 6.4**  Use the Continuity Principle to show that there is no injection from the collection of choice sequences of natural numbers into the natural numbers.[4]

In the opposite direction, an injection is easy to construct: map the number $x$ to the choice sequence that just keeps repeating it.

The Continuity Principle is moreover already sufficient to prove:

**Theorem 6.5**  (Continuity Theorem) *Every total function* $[0, 1] \to \mathbb{R}$ *is continuous.*

Brouwer never paused to state this theorem and prove it that way,[5] and instead proved the stronger Uniform Continuity Theorem (Theorem 6.10 below). For that he used, beyond the Continuity Principle, the so-called *Fan Theorem*, which we take up next.


## 6.4   The Fan Theorem and Some Applications


**Theorem 6.6**  (Fan Theorem 1) *If $W$ is a fan, then every $F : W \to \mathbb{N}$ is uniformly continuous.*

A slightly more general version is

**Theorem 6.7**  (Fan Theorem 2) *Let $A$ be a decidable property of the nodes of a fan; then* $\forall \alpha \exists x\, A(\overline{\alpha}x) \to \exists z\, \forall \alpha\, \exists y < z\, A(\overline{\alpha}y)$.

In words, if every path passes through a node that has the property $A$, then there is an upper bound on the length of the intended nodes. More informally, if every

---

[4] This argument made its first appearance in print in [21, p. 113].

[5] So see for example [83, p. 305].

**Fig. 6.4** A bar in a fan

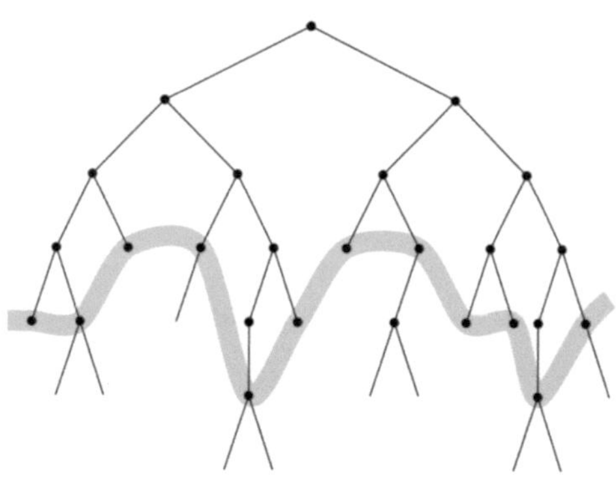

branch is cut at such a node, then there is a certain length below which all branches have already been cut. If we omit the nodes that come after a node with the property $A$, we still have a fan. According to the Fan Theorem, that truncated fan is finite. A species of nodes that "cuts off" all branches of a fan (or spread) is, using another metaphor, called a *bar* (Fig. 6.4).

We will not go into the proof of the Fan Theorem here, or the *Bar Theorem* of which it is a direct consequence. Brouwer's proof of the latter is unorthodox and fascinating, but we must refer to the literature, as it is too involved for the present book.[6] Here we will make use of the Fan Theorem without further ado.

*Remark*: in the Fan Theorem the natural numbers play the role of a species with a decidable equality, each element of which can be calculated in finitely many steps. The rational numbers would do just as well; we can see that immediately, because there is an effective coding of the rational numbers in the natural numbers.

The Fan Theorem in its second form says that the topological space given by a fan is *compact*.

**Theorem 6.8** *Every open cover of a fan has a finite subcover.*

**Proof** Let $O$ be an open cover of the fan; then for every sequence $\alpha$ there is an open $U \in O$ such that $\alpha \in U$. And so there is for every $\alpha$ an $n$ with $\alpha \in \overline{\alpha}n \subseteq U$. The Fan Theorem says that there is a maximum $n_0$ for these numbers $n$, which means that the fan is covered by the finitely many basic neighborhoods $\overline{\alpha}n$, and thus by the finitely many associated $U$'s.                                                                    $\square$

Contraposing the Fan Theorem and reading it classically yields *König's Lemma*, which however was not proved that way. It states that a fan with infinitely many nodes has an infinite branch.

König's lemma is constructively incorrect. Consider the species of all sequences $0^n = 0_1, 0_2, \ldots, 0_n$ and $1^m = 1_1, 1_2, \ldots, 1_m$, $(m, n \geq 0)$; within this we define a tree $T$ by

---

[6] E.g., [48, Sect. 3.4], [34, Sect. 3.4], [84, Chap. 4], [107, Sects. 13, 15, 17].

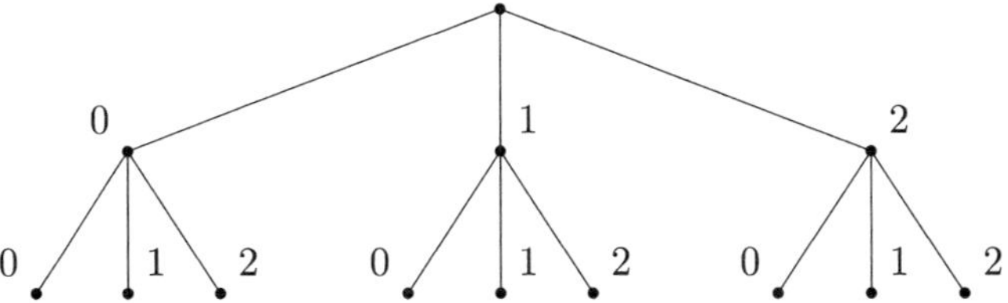

**Fig. 6.5** Top levels of the ternary fan

$$0^n \in T \leftrightarrow \forall p{<}2n \; \neg N_{20}(p) \vee \exists p{<}n \; N_{20}(2p)$$
$$1^n \in T \leftrightarrow \forall p{<}2n{+}1 \; \neg N_{20}(p) \vee \exists p{<}n \; N_{20}(2p+1)$$

In words: to test whether a sequence of zeros or ones of length $n$ is in $T$, we calculate whether there is a nine-sequence in the decimal expansion of $\pi$ to $2n + 1$. If so, see if it occurred in an even or odd place; in the first case the 1-sequence is allowed, in the second the 0-sequence. If not, $0^n$ and $1^n$ are both allowed. The tree $T$ is clearly infinite, but we cannot say whether there is an infinite branch.

The Fan Theorem has some powerful implications, some of which we will discuss here. We first prove:

**Theorem 6.9** *Closed intervals are represented by fans, that is, a closed interval is the continuous image of a fan.*

***Proof*** We start with the construction of a fan for the Cauchy sequences in $[0, 1]$, and consider a tree in which every node has exactly three direct successors: a ternary fan. The highest node, the top, is the empty sequence ( ), after which time and again the choices 0, 1, 2, are allowed. In Fig. 6.5 we have indicated only a few nodes with their values.

To the nodes we now assign rational numbers as follows:

$$\begin{cases} a_{(0)} = 2^{-1} - 2^{-2} = 1 \cdot 2^{-2} \\ a_{(1)} = 2^{-1} \qquad\quad = 2 \cdot 2^{-2} \\ a_{(2)} = 2^{-1} + 2^{-2} = 3 \cdot 2^{-2} \end{cases} \quad\text{and}\quad \begin{cases} a_{***0} = a_{***} - 2^{-k-2} \\ a_{***1} = a_{***} \\ a_{***2} = a_{***} + 2^{-k-2} \end{cases}$$

where for the sequence $***$ of length $k$ the $a_{***}$ has already been determined. See Fig. 6.6. Note that whenever we extend a sequence by choosing 1, the fraction assigned remains the same, but in the figure we rewrite it so as to have the same denominator as that of the fractions assigned to the choices 0 and 2.

The choice sequences of these $a_{****}$'s are Cauchy sequences and, in any case, all (finite) binary fractions between 0 and 1 are generated. The generated numbers all lie in $[0, 1]$. Now we have to show that every $s \in [0, 1]$ coincides with such a Cauchy sequence. For clarity, the proof we are about to give proceeds in a number of stages, and is not as short as possible.

(1) Every $s \in [0, 1]$ has a Cauchy sequence with all terms between 0 and 1. Consider an $s$ with $0 \leq s \leq 1$; we first look at $0 \leq s$. According to Theorem 4.38(4), $s = |s|$, so we can replace the terms of $(s_n)$ with their absolute values.

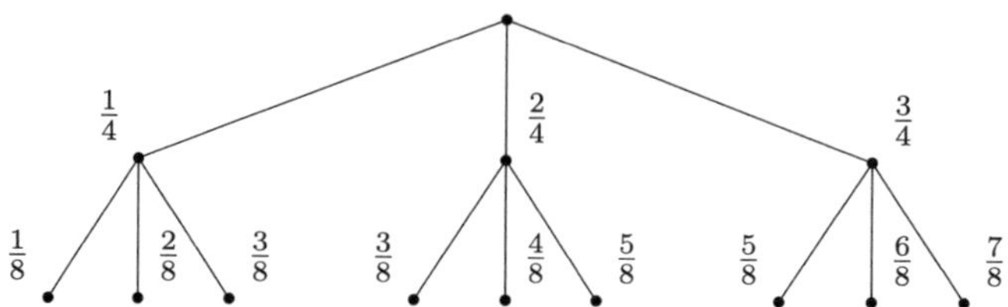

**Fig. 6.6** Top levels of the ternary fan after application of the assignment $a\_$ to the nodes

Now consider the sequence $(|s_n| + 2^{-n})$; according to Exercise 4.56, we have $(|s_n| + 2^{-n}) \sim (|s_n|) \sim (s_n)$. For $s \le 1$ we can make an analogous argument. This proves the claim.

(2) Every $s \in [0, 1]$ has a Cauchy sequence $(t_n)$ whose elements are binary fractions. The idea is to traverse the fan of Fig. 6.6 while staying as closely as possible to $s$. Convergence may be slower, but we will catch up!

*Step 0:* $0 < s_0 < 1 \rightarrow t_0 = 2^{-1}$. (We are at the root of the fan.)

*Step 1:* Choose from $1/4$, $2/4$ and $3/4$; for this we divide the interval into closed intervals with length $1/8$. Then determine $n_1$ such that $\forall m\,(|s_{n_1} - s_m| < 1/8)$. The neighborhood $(s_{n_1} - 1/8,\, s_{n_1} + 1/8)$ contains at most one of the above three points. We now define $t_1$ to be that point of the three closest to $s_{n_1}$, or in case it is exactly between two of them, the middle one. In formulas:

$$
t_1 = \begin{cases}
\frac{1}{4} & \text{if } |s_{n_1} - \frac{1}{4}| < |s_{n_1} - \frac{2}{4}| \\[4pt]
\frac{3}{4} & \text{if } |s_{n_1} - \frac{3}{4}| < |s_{n_1} - \frac{2}{4}| \\[4pt]
\frac{2}{4} & \text{else}
\end{cases}
$$

which we will rewrite as:

$$
t_1 = \begin{cases}
2^{-2} = t_0 - 2^{-2} & \text{if } |s_{n_1} - 2^{-2}| < |s_{n_1} - 2^{-1}| \\[4pt]
2^{-1} + 2^{-2} = t_0 + 2^{-2} & \text{if } |s_{n_1} - 2^{-1} - 2^{-2}| < |s_{n_1} - 2^{-1}| \\[4pt]
2^{-1} = t_0 & \text{else}
\end{cases}
$$

*Step $p + 1$:* Let $t_p$ already be determined. Determine $n_{p+1}$ such that $\forall m > n_{p+1}(|s_{n_{p+1}} - s_m| < 2^{-p-3})$.

Define:

$$
t_{p+1} = \begin{cases}
t_p - 2^{-p-2} & \text{if } |s_{n_{p+1}} - (t_p - 2^{-p-2})| < |s_{n_{p+1}} - t_p| \\[4pt]
t_p + 2^{-p-2} & \text{if } |s_{n_{p+1}} - (t_p + 2^{-p-2})| < |s_{n_{p+1}} - t_p| \\[4pt]
t_p & \text{else}
\end{cases}
$$

Note that $|t_p - s_{n_p}| < 2^{-p}$, so the sequences $(t_p)$ and $(s_{n_p})$ are equivalent, and since $(s_{n_p})$ is an infinite subsequence of $(s_n)$, we have shown that $(t_p)$ and $(s_n)$ determine the same point.

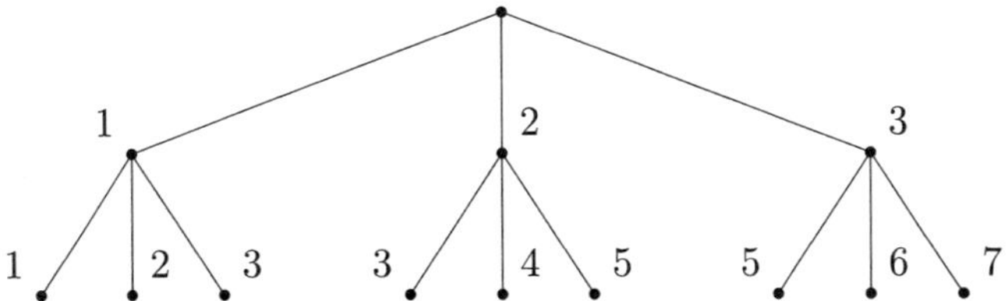

**Fig. 6.7** Top levels of the fan in Fig. 6.6 with just the numerators at the nodes, used in the proof of Theorem 6.10

If two of the Cauchy sequences defined above have an initial segment of length $n$ in common, then their distance is at most $2 \cdot 2^{-n} \cdot \sum_0^\infty 2^{-i} = 2^{-n+2}$. According to the above, the sequences passing through such an initial segment represent a closed interval of length $2^{-n+3}$.

We also note that the construction of the sequences $(s_{n_p})$ defines a continuous map of the fan on the interval $[0, 1]$. Since also intuitionistically the image of a compact species is compact, the closed interval is also compact, i.e., we have the Heine-Borel Theorem. We will also derive this result directly below as Theorem 6.16.  $\square$

The above theorem serves as a lemma to Brouwer's famous

**Theorem 6.10**  (Uniform Continuity Theorem) *Every real function on* $[0, 1]$ *is uniformly continuous.*

**Proof**  Consider the assignment $f : a \mapsto r$. By (2) in the proof of Theorem 6.9, we may consider $a$ to be given by a Cauchy sequence of the form $(a_n 2^{-n-1})$ and similarly $r$ by a sequence $(r_n 2^{-n-1})$. Let $\alpha$ be the choice sequence of the numbers $a_n$ used to construct that Cauchy sequence for $a$. For varying $a$, their associated sequences $\alpha$ are the infinite paths through another fan; see Fig. 6.7. When we now choose a fixed $n$, we can construct a function $f^* : \alpha \mapsto r_n$.

To $f^*$ we can apply the Fan Theorem: in the fan of Fig. 6.7 there is an upper bound on the length of initial segments of choice sequences that determine the same image for all their continuations. Let $k$ be such an upper bound; then every $\beta$ with the same initial segment $\overline{\alpha}k$ yields the same function value under $f^*$, $r_n$.

This means that, if $|a - b| < 2^{-k+2}$, then $|f(a) - f(b)| < 2^{-n}$, where $a$ and $b$ are the real numbers associated with $\alpha$ and $\beta$. Observe that $k$ depends on $n$, but not on $a$. This shows the uniform continuity of $f$.  $\square$

As a corollary, we at last prove that every real function on $[0, 1]$ is continuous.

**Proof of Theorem** 6.5 Directly implied by Theorem 6.10.  $\square$

We also have

**Corollary 6.11**  *Every real function on* $[a, b]$ *with* $a < b$ *is uniformly continuous.*

**Proof**  For every $a$ and $b$ with $a < b$, there is a bijective linear function $g$ with $g : [a, b] \to [0, 1]$; finish this proof yourself.  $\square$

Alternatively, we make a direct representation of such a closed interval $[a, b]$ by a fan, and imitate the proof for $[0, 1]$.

In turn, we get

**Corollary 6.12** *Every real function on $\mathbb{R}$ is continuous.*

**Proof** Let $c$ be a real number. Choose $a$ and $b$ such that $c$ lies in $[a, b]$ and $a < b$. Now $f$ is uniformly continuous on $[a, b]$ by Corollary 6.11, and hence continuous at $c$. (Of course, $f$ doesn't have to be uniformly continuous on all of $\mathbb{R}$.)    □

There is an interesting consequence of such Continuity Theorems: *the continuum is unsplittable*. A collection is *splittable* if it can be defined as the union of two disjoint, inhabited collections; the continuum can not. With obvious partitions, such as $(\infty, 0] \cup (0, \infty)$, one readily imagines that not all real numbers are included. After all, the numbers for which the location relative to 0 is indeterminate cannot yet be included in either species. But in general a clever argument is needed.

**Corollary 6.13**

$$[0, 1] = A \cup B \wedge A \cap B = \emptyset \;\rightarrow\; (A = \emptyset \vee B = \emptyset)$$
$$\mathbb{R} = A \cup B \wedge A \cap B = \emptyset \;\rightarrow\; (A = \emptyset \vee B = \emptyset)$$

**Proof** We give the proof for $[0, 1]$. Suppose $[0, 1] = A \cup B \wedge A \cap B = \emptyset$, then each $x$ in $[0, 1]$ is in $A$ or in $B$. Given the strong meaning of the intuitionistic disjunction, we can therefore give a definition by cases.

$$f(x) = \begin{cases} 0 \text{ if } x \in A \\ 1 \text{ if } x \in B \end{cases}$$

It is easy to see that there are exactly two continuous functions of $[0, 1]$ to $\{0, 1\}$, namely the constant functions $x \mapsto 0$ and $x \mapsto 1$. And that means $A = \emptyset$ or $B = \emptyset$.    □

The Continuity Theorem and the unsplittability of the continuum are examples of intuitionistic theorems that contradict  classical mathematics. "Contradict" in outward appearance, of course, for the classical mathematician and the intuitionist assign different meanings to the words or formalism used to express the theorem. Brouwer used the unsplittability of the continuum to show that a particular instance of the universally quantified predicate-logical version of PEM, under its constructive meaning, is false:

$$\neg \forall x \in \mathbb{R}\, (x \in \mathbb{Q} \vee x \notin \mathbb{Q})\,.$$

This goes beyond a weak counterexample, which only shows that as yet there is no constructive proof for a certain proposition. Instead, a "strong counterexample" such as we have at our hands here shows that there *cannot be* a proof.

Finally, we look at coverings of $\mathbb{R}$ and of closed intervals. For starters, coverings can always be replaced by open coverings. First, we define the *interior* of $X$ as the union of all open intervals contained in $X$:

**Definition 6.14**  Let $X \subseteq \mathbb{R}$, then $\text{int}(X) := \{x \mid \exists k((x - 2^{-k}, x + 2^{-k}) \subseteq X)\}$

Now we can state

**Theorem 6.15**  *If* $[0, 1] \subseteq \bigcup_0^\infty X_i$, *then also* $[0, 1] \subset \bigcup_0^\infty \text{int}(X_i)$.

***Proof***  Since the points of $[0, 1]$ are represented by choice sequences $\alpha$, we can express the covering by the $X_i$'s as follows: $\forall \alpha \exists n (x_\alpha \in X_n)$, where $x_\alpha$ is the real number associated with the sequence $\alpha$. The Continuity Principle now yields $\forall \alpha \exists m \exists n \forall \beta (\overline{\alpha} m = \overline{\beta} m \to x_\beta \in X_n)$. In other words, $H_n = \{x_\beta \mid \beta \in \overline{\alpha} m\} \subseteq X_n$. $H_n$ represents a closed interval around $x_\alpha$. Within this interval we can choose an open interval around $x_\alpha$ that is also within $X_n$. It follows that every $x_\alpha$ lies in the interior of $X_n$.

To extend the theorem to $\mathbb{R}$, use $\mathbb{R} = \bigcup_{i>0}[-i, i]$.                        $\square$

Note that this theorem shows once again that classical partitions of $\mathbb{R}$ are not intuitionistically acceptable. Suppose $\mathbb{R} = (-\infty, 0) \cup \{0\} \cup (0, \infty)$, then also $\mathbb{R} = (-\infty, 0) \cup (0, \infty)$, which is not the case. The same goes for $[0, 2] = [0, 1] \cup [1, 2]$.

**Theorem 6.16**  (Heine-Borel Theorem) *Every open cover of* $[a, b]$ *contains a finite subcover.*

***Proof***  It is enough to consider $[0, 1]$. Let $[0, 1] \subset \bigcup_{i>0} W_i$, for open $W_i$. Then for every point of $[0, 1]$, that is, for every $\alpha$, there is a $W_i$ containing $x$, that is, $x_\alpha$. According to the representation theorem, there is an $n$ such that the image of $\overline{\alpha} n$ falls within the $W_i$. Determine the maximum value of the $n$'s, as guaranteed by the Fan Theorem; this yields a finite number of covering $\overline{\alpha} n$'s, and thus a finite number of covering $W_i$'s.                                                                      $\square$

We have already established that a continuous function on a closed interval has an infimum and a supremum. Using the Fan Theorem, we can prove that a function that is positive everywhere on a closed interval has a positive infimum:

**Theorem 6.17**  *Let* $f : [a, b] \to \mathbb{R}$. *Then* $\forall x \in [a, b]\ (f(x) > 0) \to \inf(f) > 0$.

***Proof***  Since $\forall x (f(x) > 0)$, for every $x$ there is an $n$ such that $f(x) > 2^{-n}$. That is, $x \in W_n$, where $W_n = \{y \mid f(y) > 2^{-n}\}$. The collection $\{W_n \mid n \geq 0\}$ is therefore a cover. Because of the compactness of $[a, b]$, there is an $n_0$ such that $\{W_n \mid n < n_0\}$ is a cover. Now $\forall x (f(x) > 2^{-n_0})$.                                          $\square$

When Brouwer came to see that among the objects that are constructible from the basic intuition are choice sequences and spreads, and that these could be used to construct a mathematical model of the intuitive continuum, this enabled him to produce, over the years, a fine-grained theory, of which we have here shown some of the basic ideas and theorems. However, having this mathematical model did not mean that he now could do without the basic intuition. Even in his Viennese lectures of 1928, at a high point in the technical development of his program, he pointed out that intuitionism confirms the view held by Kant and Schopenhauer that the continuum is given in intuition, and that the insight that choice sequences and spreads are constructible objects itself depends on the basic intuition [28, p. 434].

## 6.5   The Brouwer-Kripke Schema

In 1927 Brouwer for the first time exploited the fact that the activity of constructing mathematical objects and proving theorems about them can itself be structured mathematically. This enabled him to give new weak and strong counterexamples. It was only some twenty years later that he published this method, which has since become known as *the method of the Creating Subject*. The theme was taken up again by Kreisel, Kripke, Myhill, Troelstra, and van Dalen. Applications of the insight were then also found beyond counterexamples.[7]

Kreisel and Troelstra proposed axioms for a theory of the Creating Subject, called CS. Without further analyzing the theory, we present one form of it here. The new basic concept added to ordinary arithmetic or analysis was an operator $\Box_n$; we read $\Box_n A$ as "at moment $n$, $A$ is evident to the subject". This evidence can be a matter of direct insight, or of recognizing that a certain chain of inferences that starts from evident premises and concludes with $A$ is correct. Note, by the way, that it is assumed that the mental acts of the Creating Subject make up a sequence ordered like that of the natural numbers in their natural order; so $n$ ranges over $\mathbb{N}$. In the light of the basic intuition, this is a plausible assumption.

The (current) axioms are

$$\Box_n A \vee \neg \Box_n A \qquad \text{(CS1)}$$
$$\Box_n A \rightarrow \Box_{n+m} A \qquad \text{(CS2)}$$
$$\exists n \Box_n A \leftrightarrow A \qquad \text{(CS3)}$$

CS1 says that at any time the subject can determine whether it has conclusive evidence for $A$. The subject knows, so to speak, what it knows. In case of doubt, $\neg \Box_n A$ applies.

CS2 states that the subject a perfect memory. That is a reasonable idealization. One can also put it this way: what the subject has once sufficiently established for itself, remains correct.

CS3, in a sense, defines what is true for the subject: that which it has at some point come to see as evident.

It is not difficult to produce, from these axioms, an infinite sequence of zeros and ones that define the state of $A$. Let us assume that the Creating Subject (into which the readers may project themselves) wants to solve a certain problem; it will be tackling the problem from all sides for a long time in all sorts of ways and—let us assume—it keeps an accurate protocol on its progress. For the sake of convenience, we will state that it notes per unit of time whether the problem has already been solved: a 0 for "not yet solved" and a 1 for "solved". This creates a choice sequence of zeros and ones.

This is a way to motivate the following principle:

$$\exists \alpha (A \leftrightarrow \exists n(\alpha(n) = 1))$$

---

[7] For an overview of such applications, see [88, pp. 1591–1592], which also provides background, discussion, and further references for this section.

This formulation comes from Kripke; in the literature it was therefore first known as *Kripke's Schema*, KS. But since the reasoning it codifies was already used informally by Brouwer, the name *Brouwer-Kripke Schema*, BKS, is more appropriate. We call the sequence $\alpha$ a *Brouwer-Kripke sequence* for $A$, BK-sequence for short.

We note that such a sequence can still have various degrees of freedom. For example, it is possible to agree that a 1 will occur at most once, or that a 1 may only be followed by 1's. It doesn't even have to be a 0–1 sequence. When using BKS, one can always choose some convenient form.

While the three axioms CS1-3 were formulated after BKS, they make explicit the ideas used in the justification of BKS.

In combination with earlier insights, Brouwer in effect used BKS to produce a result that, although it seems plausible, had so far eluded proof. One may well suspect that $a \# b$ is stronger than $a \neq b$, but an example of two unequal numbers that are not apart cannot be given because $\neg a \# b \leftrightarrow a = b$. Brouwer found that nevertheless something can be said about the difference between $\#$ and $\neq$ if we look at all real numbers together:

**Theorem 6.18**  $\neg \forall x \in \mathbb{R}\ (x \neq 0 \rightarrow x \# 0)$

***Proof*** Assume $\forall x (x \neq 0 \rightarrow x \# 0)$.

We define a BK-sequence $\alpha$ which depends on $r \in \mathbb{Q} \vee r \notin \mathbb{Q}$,

$$\exists x (\alpha(x) = 1) \leftrightarrow (r \in \mathbb{Q} \vee r \notin \mathbb{Q})$$

We now define a real number $a$ by a Cauchy sequence :

$$a_n = \begin{cases} 2^{-n} \text{ if } \forall k \leq n(\alpha(k) = 0) \\ 2^{-k} \text{ if } (\alpha(k) = 1 \wedge \forall p < k(\alpha(p) = 0) \text{ and } k \leq n \end{cases}$$

From the definition it follows that $a = 0 \rightarrow \forall n(\alpha(n) = 0) \rightarrow \neg(r \in \mathbb{Q} \vee r \notin \mathbb{Q}) \rightarrow \neg(r \in \mathbb{Q}) \wedge \neg\neg(r \in \mathbb{Q}) \rightarrow$ contradiction. So $a \neq 0$. It follows from the assumption that $a \# 0$, which means that $\exists k(a_n > 2^{-k})$, which means by definition that $\exists k(\alpha(k) = 1)$. Therefore, $r \in \mathbb{Q} \vee r \notin \mathbb{Q}$ applies. Since $r$ was an arbitrary real number, we get $\forall x \in \mathbb{R}\ (x \in \mathbb{Q} \vee x \notin \mathbb{Q})$. And this contradicts the unsplittability of the continuum. $\qquad\square$

So we have a strong counterexample to the claim that inequality and apartness are equivalent. (They are equivalent in the different setting of Markov's Recursive Mathematics; see Theorem 6.28.)

Note that the strength of the proof lies in the combination of the Continuity Principle (via the unsplittability of the continuum) and BKS.

Whereas this last result was not unexpected, BKS also has consequences that are at odds with common expectations. Here are two cases.

(i) The first has to do with the concept of species: one would like to know whether the power species of a set is an admissible concept. Our first experience with the power species goes back to the finite collections. Here the situation, at least classically, is fairly straightforward, as one can enumerate the subsets: for example, $\mathcal{P}\{0, 1, 2\} = \{\emptyset, \{0\}, \{1\}, \{2\}, \{0, 1\}, \{0, 2\}, \{1, 2\}, \{0, 1, 2\}\}$. Unfortunately, this is of little help to an intuitionist, for whom there are many more subspecies than those listed above. Here is an example of an overlooked subspecies: $\{n < 3 \mid (n = 0 \wedge R) \vee (n = 2 \wedge \neg R\}$, where for $R$ we take the as yet undecided Riemann Hypothesis. One can make up arbitrarily many such species.

Even the singleton species $\{0\}$ already has infinitely many subspecies—each mathematical proposition $A$ can be used to define a new subspecies $\{n < 1 \mid A\}$.

The question we have not answered is whether the power species of $\{0\}$ exists, or of $\mathbb{N}$. If we cling to the view that we only recognize collections that are constructed from a previously given totality, then we are in trouble. There is little hope of indicating all sets of natural numbers. But we surely can allow collections consisting of elements of spreads, i.e., infinite sequences of objects, since spreads are the next step in the intuitionistic universe, after the countable species. Let us see if this leads anywhere.

Consider subspecies $X$ and elements $n$ of $\mathbb{N}$. Apply BKS to the statement $n \in X$:

$$\forall X \; \forall n \; \exists \alpha \in \{0, 1\}^{\mathbb{N}} \; (n \in X \leftrightarrow \exists m (\alpha(m) = 1))$$

This says that for every $n$ there is a sequence $\alpha$ (i.e., a function from $\mathbb{N}$ to $\{0, 1\}$ such that ...) We can combine all these functions together into a single function using an encoding of pairs of natural numbers into natural numbers. Let $f(x, y)$ be an encoding of $\mathbb{N} \times \mathbb{N}$ to $\mathbb{N}$. We now apply the Axiom of Countable Choice in the form $\forall n \exists \alpha R(n, \alpha) \rightarrow \exists \alpha \forall n R(n, \alpha_n)$, where $\alpha_n(m) = \alpha(f(n, m))$. The result is

$$\forall X \exists \alpha \forall n (n \in X \leftrightarrow \exists m (\alpha(f(n, m)) = 1))) ,$$

which can be read as "any subspecies $X$ of $\mathbb{N}$ is represented by a choice sequence $\alpha$, as indicated in the formula". So each collection yields a 0-1 function on $\mathbb{N} \times \mathbb{N}$, and conversely every 0–1-function on $\mathbb{N} \times \mathbb{N}$ yields, in the manner of BKS, a collection. By encoding $\mathbb{N} \times \mathbb{N}$ on $\mathbb{N}$ each branch in the binary tree corresponds to a subspecies from $\mathbb{N}$. Ergo, $\mathcal{P}(\mathbb{N})$ is represented by a fan, making this power species acceptable by Brouwerian standards. See further [94].

(ii) Brouwer's Continuity Principle teaches us that functions from $\mathbb{N}^{\mathbb{N}}$ to $\mathbb{N}$ are continuous. Isn't it obvious to go one step further and assume that functions of $\mathbb{N}^{\mathbb{N}}$ to $\mathbb{N}^{\mathbb{N}}$ are also continuous? After all, if $f(\alpha) = \beta$, then increasingly longer initial segments of $\beta$ are calculated by using longer and longer initial segments of $\alpha$ in the calculation. We call this extended continuity $\forall \alpha \exists \beta$-*continuity*. Unfortunately, BKS shows us that things are a little more complicated.

**Theorem 6.19** *BKS and* $\forall \alpha \exists \beta$-*continuity conflict [65].*

***Proof*** Towards a contradiction, we assume that both principles are true. Apply BKS to $\forall x (\alpha(x) = 0)$:

$$\exists \beta (\forall x (\alpha(x) = 0 \leftrightarrow \exists y (\beta(y) = 1))$$

This shows: $\forall\alpha\exists\beta(\ldots)$; so by $\forall\alpha\exists\beta$-continuity there must be a continuous $G\colon \mathbb{N}^{\mathbb{N}} \to \mathbb{N}^{\mathbb{N}}$ such that $\forall\alpha((\forall x(\alpha(x) = 0 \leftrightarrow \exists y(G(\alpha)(y) = 1))$. That means that we have a continuous function $G$ that tests whether $\alpha$ is the zero sequence $\mathbf{0}$, that is, $G$ is $\mathbf{0}$ on all sequences other than $\mathbf{0}$, and not equal to $\mathbf{0}$ on $\mathbf{0}$. This function is clearly discontinuous. Suppose $\beta = G(\mathbf{0})$ and $\beta(n) = 1$. Now consider the neighborhood $O = \overline{\beta}(n+1))$. Because of the continuity of $G$, there is a neighborhood $O'$ of $\mathbf{0}$, say $\overline{\mathbf{0}}(m)$ that is mapped entirely within $O$. That is impossible since the sequences through $\overline{\mathbf{0}}(m-1)$ are mapped by $G$ to sequences that fall within $O$, and thus take the value 1 on $n$.   $\square$

For the intuitionist there is a choice here: accept either BKS, or $\forall\alpha\exists\beta$-continuity, or neither; but not both. There are in fact good reasons for accepting BKS. As noted, BKS follows immediately from Brouwer's ideas about the Creating Subject, and there are several places in Brouwer where he reasons along its lines. In contrast, this is not the case for $\forall\alpha\exists\beta$-continuity, to which Brouwer moreover in effect constructed a strong counterexample. For the details and discussion, see [88].

A last example: BKS contradicts the Church-Turing Thesis (CT), which identifies the informal concept of calculability with the formal one of recursiveness. (For more on CT, see Sect. 6.9.1.) We will give the argument using the axioms CS instead, as that simplifies the presentation. For two versions of the argument with BKS, see [88, Sect. 7.6].

*Notation*: The complement of a collection $X$ is denoted by $X^c$. Usually that is understood to be relative not to "everything" (a difficult concept), but to a more specific collection given in the context.

Let $K$ be a species of natural numbers that is recursively enumerable, but its complement $K^c$ (i.e., $\mathbb{N} - K$) is not.[8] Define

$$f(n, m) = \begin{cases} 0 & \text{if } \neg\square_m n \notin K \\ 1 & \text{if } \square_m n \notin K \end{cases}$$

Then

$$n \notin K \leftrightarrow \exists m f(n, m) = 1$$

From left to right, this follows from CS3 read from right to left; from right to left, this follows from CS3 read from left to right. For the Creating Subject this function $f$ is calculable, as, by CS1, for any given $m$ and $n$, $\square_m n \notin K$ is decidable. Assume that $f$ is moreover recursive. Then the species $S = \{n \mid \exists m f(n, m) = 1\}$ is recursively enumerable, but as $S$ is equivalent to $K^c$, this contradicts the hypothesis.

There is no reason to think that no further principles like BKS can be found that give intuitionistic mathematics further strength beyond general constructive mathematics.

---

[8] Example: The set of code numbers of Turing Machines that halt when given their own code number as input.

## 6.6   More About Discontinuity

In the foregoing we have shown that in Brouwer's mathematics the continuity of real functions is a consequence of the basic principles associated with choice sequences. In general, even without these principles, there is still something to be said about discontinuity. The key property here is the undecidability of the equality of (lawlike or non-lawlike) choice sequences.

In the following we consider the tree of all finite sequences of natural numbers and the associated spread $\mathbb{N}^{\mathbb{N}}$ of choice sequences. On this spread let a function $f$ be given with range $\mathbb{N}$. We assume that $f$ is discontinuous; without loss of generality, we can assume that $f$ is discontinuous at the zero sequence, $\mathbf{0}$, and that $f(\mathbf{0}) = 0$. Geometrically, the discontinuity at $\mathbf{0}$ means that in the tree branches keep splitting off from $\mathbf{0}$ that get a positive value under $f$:

$$\forall n \exists \alpha \in \overline{\mathbf{0}}n \ (f(\alpha) \neq 0) . \tag{6.1}$$

Note that the split itself of such an $\alpha$ from $\mathbf{0}$ need not occur immediately after $\bar{\alpha}n$, but may happen much later. Because of the quantifier combination $\forall n \exists \alpha$, we can construct a sequence of such $\alpha$'s: $\alpha_0, \alpha_1, \alpha_2, \ldots$. We assume that every $\alpha_{n+1}$ splits off lower in the tree than $\alpha_n$; this is no restriction. (Why not?) Now, it need not be the case that $\alpha_{n+1}$ is the *first* branch lower in the tree than $\alpha_n$ that splits off from $\mathbf{0}$. This is because the existential quantifier in (6.1) only asks for *some* sequence $\alpha$ satisfying the stated condition, but requires no more from it. So between an $\alpha_n$ and $\alpha_{n+1}$, the spread may well contain other branches splitting off from $\mathbf{0}$.

We now create the smaller spread $S$ that is formed by just the nodes of the sequences $\mathbf{0}, \alpha_0, \alpha_1, \alpha_2, \ldots$. We define a map $g$ from $\mathbb{N}^{\mathbb{N}}$ to $S$ by sweeping all branches of the large spread leftward onto the small spread (see Fig. 6.8). That is, for every $\alpha$ we determine the one $n$ such that $\alpha$ splits off from $\mathbf{0}$ after $\alpha_{n-1}$ and before or at $\alpha_n$. So $g : \alpha \mapsto \alpha_n$.

**Fig. 6.8** The mapping $g : \mathbb{N}^{\mathbb{N}} \to S$ sweeps $\alpha$ onto $\alpha_3$

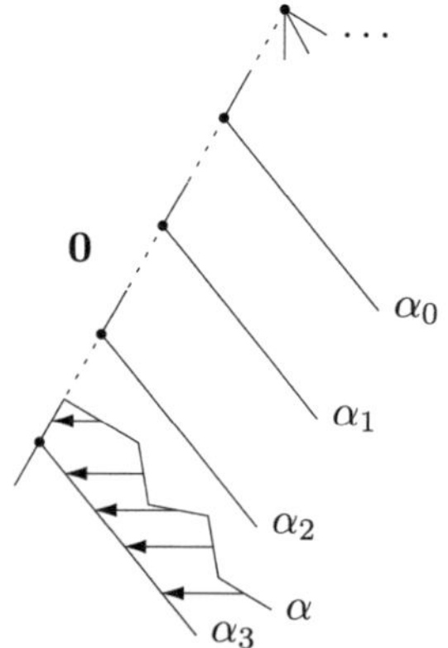

From the definition of $g$ it follows that

$$\forall \alpha (\alpha \ \# \ \mathbf{0} \leftrightarrow f(g(\alpha)) \neq 0) \,,$$

so

$$\forall \alpha (\alpha = \mathbf{0} \leftrightarrow f(g(\alpha)) = 0) \,.$$

Since equality of natural numbers is decidable, we have obtained a test for equality of choice sequences with $\mathbf{0}$: $\forall \alpha (\alpha = \mathbf{0} \vee \alpha \neq \mathbf{0})$. But weak counterexamples show that this property cannot be defended at all.

The similarity test can be extended to arbitrary sequences without further ado:

$$\forall \alpha \forall \beta (\alpha = \beta \vee \alpha \neq \beta) \,.$$

The converse is immediately apparent: if the equality between choice sequences is decidable, then discontinuous functions exist. Conclusion:

**Theorem 6.20**  *There are discontinuous* $f : \mathbb{N}^{\mathbb{N}} \to \mathbb{N} \leftrightarrow \forall \alpha \forall \beta (\alpha = \beta \vee \alpha \neq \beta)$.

Assuming that all sequences $\alpha$ are recursive (CT), there are no discontinuous functions of $\mathbb{N}^{\mathbb{N}}$ on $\mathbb{N}$. After all, recursion theory has proven that the equality between recursive functions is undecidable.

**Corollary 6.21**  *Assuming CT, there are no discontinuous functions from* $\mathbb{N}^{\mathbb{N}}$ *to* $\mathbb{N}$.

Let us now approach the problem of discontinuous functions in relation to BKS, and strongly extensional functions (see Definition 5.2). We again use the formulation of BKS for sequences of zeros and ones, with at most one 1. We repeat the construction from the previous theorem for $f$.

**Theorem 6.22**  *Assuming BKS, the existence of strongly extensional discontinuous functions is equivalent to PEM.*

***Proof***  Take an arbitrary statement $P$ and apply BKS to $P \vee \neg P$. We call the corresponding BK-sequence $\alpha$:

$$\exists x (\alpha(x) = 1) \leftrightarrow P \vee \neg P$$

Now define the sequence $\delta$ by

$$\delta(n) = \begin{cases} 0 \text{ if } \forall k \leq n(\alpha(k) = 0) \\ \alpha_k(n) \text{ if } k \leq n \text{ and } \alpha(k) = 1 \end{cases}$$

using the $\alpha_k$'s introduced above.

For a function $f : \mathbb{N}^{\mathbb{N}} \to \mathbb{N}$ we then determine $f(\delta)$. Now $f(\delta) = 0 \vee f(\delta) \neq 0$. If $f(\delta) = 0$, then $\forall n(\alpha(n) = 0)$, that is, $\neg(P \vee \neg P)$. Contradiction.

If $f(\delta) \neq 0$, then we get $\delta \mathbin{\#} 0$, since the inequality for the natural numbers is an apartness relation, and $f$ is strongly extensional. But that means that $\delta = \alpha_n$ for a certain $n$. Therefore, $\exists x (\alpha(x) = 1)$ holds, and therefore also $P \vee \neg P$.

Since $P$ was an arbitrary statement, we are led to conclude to the validity of the schema $P \vee \neg P$. $\qquad\square$

So if one accepts BKS, this yields a weak counterexample to the claim that there are such functions.

## 6.7   More About the Unsplittability of the Continuum

The specific character of the intuitionistic continuum, compared with its classical counterpart, we have already demonstrated in the Unsplittability Theorem (corollary 6.13). The continuum appears to be strongly connected. It cannot be pulled apart without losing points. In fact, the continuum is even more viscous than the Unsplittability Theorem suggests. One might think that the continuum with a perforation in it would be splittable. The following theorem show that this is not the case. Even if we shoot all rational numbers out of $\mathbb{R}$, the result is still unsplittable. A worse perforation is hardly imaginable! The theorem in fact is more general and indicates a large class of unsplittable subspecies of $\mathbb{R}$.

**Theorem 6.23**  *(BKS) Every negative, dense subspecies of $\mathbb{R}$ is unsplittable [95].*

A subspecies $X$ is called *negative* if $X = X^{cc}$. We first prove two lemmas.

**Lemma 6.24**  (BKS)  *If $X$ is a negative subspecies of $\mathbb{R}$ and $X = A \cup (X - A)$, then converging sequences in $A$ and $X - A(= A^c)$ have different limits.*

**Proof**  Let $a_i \in A$, $b_i \in A^c$ and let $(a_i)_i$ converge. Assume that
$$\forall k \exists n \forall m (|a_{n+m} - b_{n+m}| < 2^{-k})$$
We now use BKS under the assumption that the BK-sequences take the value 1 at most once.

Let $\alpha$ be a BK-sequence for $r \in \mathbb{Q}$ (where $r \in \mathbb{R}$) and $\beta$ for $r \notin \mathbb{Q}$.
Define
$$\begin{cases} \gamma(2x) & = \alpha(x) \\ \gamma(2x + 1) & = \beta(x) \end{cases}$$

$$\begin{cases} c_{2n} & = a_n \\ c_{2n+1} & = b_n \end{cases}$$

and then

$$d_n = \begin{cases} c_n \text{ if } \forall k \leq n (\gamma(k) = 0) \\ c_k \text{ if } k \leq n \text{ and } \gamma(k) = 1 \end{cases}$$

Note that $(d_n)_n$ converges, say $\lim(d_n) = d$.

*Claim:* $d \in X$. For if $d \notin X$ then $d \notin A$ holds, so $\forall n(\alpha(n) = 0)$ and therefore $r \notin \mathbb{Q}$; moreover $d \notin A^c$ holds, so $\forall n(\beta(n) = 0)$, and therefore $r \notin \mathbb{Q}^c$. Contradiction. Since $X$ is negative, we now conclude $d \in X$.

We therefore get $d \in A \vee d \in A^c$, and so $r \notin \mathbb{Q}^c \vee r \notin \mathbb{Q}$. This again produces a contradiction. This shows that $(a_i)_i$ and $(b_i)_i$ cannot have the same limit. $\qquad\square$

**Lemma 6.25** *If the negative subspecies $X$ of $\mathbb{R}$ is the disjoint union of two inhabited species $A$ and $A^c$, then there are sequences in $A$ and $A^c$ with the same limit.*

**Proof** Since $X$ is closed, it is not a limitation to assume that there is $a_0 \in A$ and $b_0 \in A^c$, such that $u = b_0 - a_0 > 0$.

We simultaneously define $a_n, b_n$.

Assume $a_0 \ldots, a_{n-1} \in A$ and $b_0, \ldots, b_{n-1} \in A^c$ are already defined, and that $|a_i - b_i| < (2/3)^i u$ for $i < n$.

Divide the interval $[a_{n-1}, b_{n-1}]$ into three intervals of equal length and choose an $x$ in the middle interval.

$$a_n = \begin{cases} x & \text{if } x \in A \\ a_{n-1} & \text{else} \end{cases}$$

and

$$b_n = \begin{cases} x & \text{if } x \notin A \\ b_{n-1} & \text{else} \end{cases}$$

Then

$$|b_n - a_n| < (2/3)^n u,$$
$$|b_n - b_{n-1}| < (2/3)^n u, \text{ and}$$
$$a_{n-1} - a_n < (2/3)^n u .$$

Hence, $(a_n)$ and $(b_n)$ converge and $\lim(a_n) = \lim(b_n)$. $\qquad\square$

**Proof of Theorem** 6.23 Immediate from the two lemmas. $\qquad\square$

There are plenty of examples of negative subspecies. Spectacular examples are the continuum with a hole in it, $\mathbb{R} \setminus \{0\}$; the irrational numbers $\mathbb{Q}^c = \mathbb{R} \setminus \mathbb{Q}$; and the non-non-rational numbers $\mathbb{Q}^{cc} = \mathbb{R} \setminus \mathbb{Q}^c$. That the corresponding sets are actually splittable in classical mathematics is not a mathematical, but a logical fact—you can do a lot with PEM. But what is really surprising is the observation that now $\mathbb{Q}^c$ and $\mathbb{Q}^{cc}$, according to the common definitions, are topologically connected; so in a sense $\mathbb{Q}^{cc}$ seems much richer than $\mathbb{Q}$. One has to be a little careful with the irrational numbers; there are also *strongly irrational numbers*, that is, irrational numbers that are apart from any rational number. Those strongly irrational numbers are given by infinite continued fractions, and form a species that *can* be split. The foregoing shows that surprises can be experienced in the intuitionistic universe. The newcomer will soon be inclined to dismiss these phenomena as "pathological", but numerous pathological phenomena have been noted in the history of mathematics and over time been accepted as "normal". One is reminded here of the case of continuous functions that have no tangent at any point. It is probably best to view these new phenomena as an unexpected wealth of mathematics. In this respect both intuitionistic and recursive mathematics differ from Bishop's.

## 6.8  More About the Axiom of Choice

In Sect. 4.2 we accepted a justification for a version of the Axiom of Choice that we used later on, AC-NN. Here we will present a counterexample to another version, and then a very simple one to the Axiom in its full generality.

The version in question is what we will, for this occasion, baptize AC-RN:

$$\forall x \in \mathbb{R} \; \exists n \in \mathbb{N} \; A(x, n) \; \rightarrow \; \exists f \in \mathbb{N}^{\mathbb{R}} \; \forall x \in \mathbb{R} \; A(x, f(x)) \tag{6.2}$$

For the strong counterexample, we begin by considering the statement

$$\forall x \in \mathbb{R} \; \exists y \in \mathbb{N} \; (x < y) \tag{6.3}$$

This is an evidently true statement, which we demonstrate as follows. Let a proof be given of a proposition of the form $x \in \mathbb{R}$; this proof takes the form of showing that some sequence $a_i$ is a Cauchy sequence, which we then take as a representative. By Definition 4.1, we can therefore construct an $n$ such that $\forall m(|a_n - a_{n+m}| < 2^0)$. Then the first natural number greater than $a_n + 1$ is such a $y$ as sought.

Now assume that AC-RN is correct. Its application to (6.3) yields a function $f$ from the real to the natural numbers such that $x < f(x)$. The only continuous functions from $\mathbb{R}$ to $\mathbb{N}$ are the constant functions, so $f$ cannot be one of them, as then it would be false that $x < f(x)$ for all $x$. So $f$ is discontinuous. But that is impossible, by the Continuity Theorem (Theorem 6.5). So AC-RN leads to a contradiction.[9]

In other varieties of constructivism, similar results can be used. In Markov's Recursive mathematics (Sect. 6.9.1), the Kreisel-Lacombe-Shoenfield-Tseitin Theorem likewise states that all functions $\mathbb{R} \to \mathbb{R}$ are continuous. And while in Bishop's constructivism (Sect. 6.9.2), strong counterexamples to classical principles cannot exist, we can devise a weak counterexample using Theorem 6.20: the existence of a discontinuous function $\mathbb{N}^{\mathbb{N}} \to \mathbb{N}$ is equivalent to decidability of equality on the real numbers, which is hopeless.

A conclusion to draw from this is that the method we just used in proving (6.3) does not embody a function. As was alluded to at the beginning of Sect. 5.2, functions are understood to be *extensional*, that is, to satisfy

---

[9] This argument is due to Dana Scott.

$$a = b \rightarrow g(a) = g(b) \tag{6.4}$$

But the same real number may be represented by rather different Cauchy sequences, and that has an effect here. Let $a$ be 4, considered as a real number, represented by a Cauchy sequence approximating it wholly from below. Let $b$ also be 4, considered as a real number, represented by Cauchy sequence approximating it wholly from above. Then this method assigns 5 to $a$, and 6 to $b$. But while the natural numbers 5 and 6 stand in the required relation to the real number 4, the result of the method is hereby shown to depend on the chosen representative, and so the method does not satisfy condition (6.4). Such methods are called *operations* instead of functions.[10]

Let us compare the situation with that of AC-NN, which we discussed in Sect. 4.2. In contrast to real numbers, natural numbers are not constructed via an equivalence relation on other objects, but as repeated applications of the successor function. When a natural number is given as a construction using other functions, say addition or multiplication, doing the arithmetic consists precisely in bringing it to that canonical form. So the whole situation where a difference between representatives is exploited to construct a counterexample to a choice principle cannot arise, and condition (6.4) is met. This explains why the justification of AC-NN given there depends not only on the Proof Interpretation, but also on the fact that the domain of the choice function is $\mathbb{N}$.

Counterexamples to specific versions of the axiom of choice will also be counterexamples to the general form. But a direct counterexample to the latter can already be given in a very simple constructive setting, without invoking analysis:

**Theorem 6.26**  (Diaconescu's Theorem) AC $\rightarrow$ PEM.

The following elegant proof (in the setting of basic constructive set theory) of Radu Diaconescu's theorem is due to Nicholas Goodman and John Myhill [42].

***Proof*** Let $A$ be an arbitrary statement. Consider the sets

$$U = \{n \in \mathbb{N} \mid n = 0 \vee (n = 1 \wedge A)\},$$
$$V = \{n \in \mathbb{N} \mid n = 1 \vee (n = 0 \wedge A)\}.$$

It is clear that $\forall X \in \{U, V\}\ \exists y \in \mathbb{N}\ (y \in X)$. A choice function $f$ for this case satisfies $\forall X \in \{U, V\}\ (f(X) \in X)$. Because $f(X)$ is a natural number and equality on the natural numbers is decidable, $f(U) = f(V) \vee f(U) \neq f(V)$. We now transform this disjunction into the one we are looking for:

(a) If $f(U) = f(V)$, then both are 0 or both are 1, and either way $A$ follows, from which we infer $A \vee \neg A$.
(b) If $f(U) \neq f(V)$, assume $A$. Then $U = \{0, 1\}$ and $V = \{0, 1\}$, so $U = V$, hence $f(U) = f(V)$, contradiction. So $\neg A$, from which we infer $A \vee \neg A$.

Therefore, $A \vee \neg A$; and as $A$ was arbitrary, PEM.                                        $\square$

---

[10] For further discussion of this distinction, see [5, pp. 40–42]

In (b), notions of extensionality come in at two places: when concluding to $U = V$ (extensionality of equality on sets or species), and when concluding from that to $f(U) = f(V)$ (extensionality of functions). On the other hand, the intensional difference between $U$ and $V$—the fact that they are defined differently—gets no chance to play a role, in contrast to what we saw in the counterargument to AC-RN above.

Diaconescu's Theorem is remarkable because here a purely mathematical principle implies a logical principle. In fact, there are many such mathematical statements. In classical mathematics, logic is, in a sense, unassailable, because it is taken to underlie mathematics. But in systems with weaker logics, there is a large latitude for interaction between mathematics and logic.

## 6.9  Intuitionism Compared with Two Other Currents in Constructive Mathematics

### 6.9.1  *Recursive Mathematics and Russian Constructivity*

Back in the 1930s, a major development grew out of Kurt Gödel's famous Incompleteness Theorems, namely, Recursion Theory, also known as the Theory of Recursive Functions and the Theory of Effective Computability. Various analyses of the nature of computation, most notably by Alonzo Church, Alan Turing, and, in a strictly numerical setting, Gödel, resulted in a stable formal notion of computability: their analyses turned out to be equivalent. This equivalence is taken by many to be an indication that the concept has been determined correctly.

The most appealing formulation comes from Turing, who introduced an abstract machine—the *Turing Machine*—which can move back and forth along an infinitely long tape, writing and reading symbols on it. Church and Turing in particular have provided strong arguments as to why *computable, algorithmic, decidable* and the like have indeed been given an absolute characterization by Turing Machines (or their equivalent formulations). The Church-Turing Thesis says that "algorithm $\approx$ Turing Machine", in the precise sense that any algorithm can be simulated by a Turing Machine. So here the informal concept of algorithm, which enables you "to recognize it when you see it", is linked to a formal mathematical concept. This kind of formalization of intuitive concepts is a very common procedure, for example think of "speed $\approx$ derivative".

The original formalisms, the applications of which could be quite intricate, were almost immediately replaced by a more-or-less axiomatic development with a convenient notation by Church's student Stephen Kleene. One place where one would expect such application would be in intuitionistic mathematics. After all, if a function is computable, it ought to be deemed constructive; and, on the Church-Turing Thesis, if a function is constructive, it has a computable algorithm underlying it. Nonetheless, no real development of intuitionistic theory along recursion-theoretic lines took place. Probably the reason for this was timing. The initial analyses of computability were published in 1936 and within a few short years, the Second World War would break out.

Once the notion of algorithm was mathematically established, one could proceed to characterize the real numbers. The obvious solution is "a constructive real number is given by a Cauchy sequence which is given by a Turing Machine". The naming in the literature is perhaps not quite what one would expect—something like "the computable or algorithmic continuum"—but the predicate that has won in the field of computability is that of *recursive*. We speak of the recursive continuum, of recursive functions, recursive sets, etc.; and also "recursive analysis".[11]

Now it makes sense to have Turing Machines perform the operations on the recursive continuum as well. For the basic operations of arithmetic, that is pretty obvious. In general, it limits the class of functions on the continuum. However, it can be proved that all real functions are continuous. This is the theorem of Kreisel-Lacombe-Shoenfield-Tseitin (KLST), proved in 1959.[12] (While this result reads the same as the corresponding one in intuitionistic mathematics, the underlying understandings of what a real number is are different, and so are the proofs.) By that time, in Russia, after the end of its Great Patriotic War, Markov had turned to recursive mathematics; we turn to his work below.

While recursive mathematics shares the general constructive outlook with intuitionistic mathematics, there turned out to be striking differences. For example, in recursive analysis one can show that on a closed interval there are continuous functions that are positive everywhere, while the infimum is still 0; in intuitionistic mathematics such a function has a positive infimum.[13]

Some people choose to identify effectiveness with recursiveness, but that does not altogether end the discussion around effectiveness. In particular, the question "Is this a successful calculation?" can be answered in different ways. Here is a somewhat pathological example: there is a Turing Machine that outputs 1 at input 0 if the Riemann Hypothesis is true, and 0 otherwise. How is this possible? The solution is surprisingly simple: make two Turing Machines, the first giving output 1 for each input and the other giving output 0 for each input. Now apply PEM: the Riemann Hypothesis is correct or incorrect, in the first case the first machine is the requested one, in the second case the second machine is the requested one.[14] That we don't know which machine is the desired one is, according to classical logic, irrelevant.

A more serious disagreement (because it is one among constructivists themselves) is this: when is a calculation, say of a Turing Machine, successful—when after a finite time the calculation ends, or when it is impossible that the machine never stops?

---

[11] About the terminology, see [75].

[12] We take this occasion to note that the particular proof of it presented in the original Dutch edition of the present book, a proof first published in [92], is not correct. Specifically, in the proof of lemma 3.5 on p. 345 of the latter, $b = n$ does not imply $\beta = \gamma$. Helpful remarks by Wim Veldman and Wu Jinyue led us to identify this.

[13] For a detailed treatment of this example and its relation to different forms of constructivism, see [15, Chap. 6].

[14] This example is attributed to Leon Henkin.

The first answer is the intuitionistic position. A calculation consists of discrete steps; through a suitable coding, the entire calculation can be reduced to effective basic steps applied to natural numbers (whoever wants to can just as well think of symbol manipulations on a tape), with the answer (output) being a natural number. That natural number is an object that is built up in finitely many steps, so the machine has to give an answer in a finite time, i.e., stop calculating. According to the intuitionistic conception of existence, the statement "the machine stops after a finite number of steps" is an abbreviated version of something like "the machine stops after 200,745,174 steps (at least a concrete number)". If necessary, an upper limit to the number of steps is sufficient: "the machine stops after a maximum of $9^{9^9}$ steps".

The second answer is more liberal. The idea here is that one turns on the machine, goes home to do other things, and now and then returns to see if the machine has already stopped—as it would seem it must, as it is impossible that it calculates forever. The constructive content of that computable concept is weaker than the original intuitionistic concept. The promise "After $n$ steps the answer is ready", is replaced by "There cannot come no answer, but when the answer will come I don't know". Note that this depends on the ability to list all the natural numbers one after another in such a way that each natural number will appear after a finite number of steps.

That liberal view stems from Markov [51, p. 18 ff], and is succinctly formulated as follows:

$$\forall x (A(x) \lor \neg A(x)) \rightarrow (\neg\neg \exists x A(x) \rightarrow \exists x A(x)). \tag{MP}$$

This is called *Markov's Principle*, or *The Principle of Constructive Choice*. It provides a form of double negation elimination, not on logical grounds, but, as we just saw, for a reason to do with the nature of the natural number sequence. MP is therefore considered an arithmetical axiom. From an intuitionistic perspective, MP amounts to a weakening of the existential quantifier for decidable predicates about natural numbers, as MP claims that in that case knowing $\neg\neg \exists x A(x)$ is already enough to conclude to $\exists x A(x)$.

A different version of MP is formulated using an equality containing a recursive function variable:

$$\neg\neg \exists x (\alpha(x) = 0) \rightarrow \exists x (\alpha(x) = 0.$$

The idea is that, if the predicate $A(x)$ above is recursively decidable, one can define a recursive characteristic function[15]:

$$\alpha(x) = \begin{cases} 0 & \text{if } A(x) \\ 1 & \text{if } \neg A(x) \end{cases}$$

---

[15] To obtain that function, the Axiom of Choice is used here in the form known as "AC-NN!". That is axiom (4.1) with unique existential quantification in the antecedent. (For each $x$, there is one and only one outcome of the decision procedure.) Hence, in settings where AC-NN! is not available, the two versions of MP are not equivalent. For (much) more on this and related matters, see [32].

This yields $\forall x(A(x) \leftrightarrow \alpha(x) = 0)$.

We step back for a moment to take a wider view on Markov and his situation.

The mathematics developed in a given time and place is influenced, possibly even governed, by the local general culture. As we noted in Chap. 1, by the beginning of the 20th century, a certain unease among mathematicians had become evident to anyone who cared to notice. Mathematics had drifted away from intuition and visible content, presenting itself as a bizarre landscape of monsters (nowhere differentiable functions, space-filling curves, etc.) and paradoxes (the set of all sets not elements of themselves, etc.). In a country dominated by ideological purity, as was Russia in the 1920s, such a "meaningless abstraction" would easily run afoul of the dominant Marxist dialectic-materialist theory. Several major Russian mathematicians thought that intuitionism, with its emphasis on construction and meaning, could be shown directly compatible with Marxist thought (see [108]).

At the time, direct connections between Brouwer and Russian mathematicians were already in place thanks to Brouwer's role as a leading topologist and his mentoring of two young men who were becoming leaders in the field themselves, Pavel Alexandrov and Pavel Urysohn. Also, by the middle of the decade, Hilbert's animosity towards intuitionism and his program to justify abstract mathematics using intuitionistic methods were widely discussed. Thus, in the 1925–1926 winter semester, the famous probability theorist Aleksandr Khinchin spoke in a special colloquium on the "Ideas of intuitionism and the struggle for a subject matter in contemporary mathematics". The aim of the colloquium was to "create working relations between Marxist theorists and prominent Russian mathematicians" [108, p. 370]. The gist of Khinchin's paper was that intuitionism had meaning and that Hilbert's program, upon completion, would reduce the abstract modern mathematics to the intuitionistic, thereby endowing it with Marxist acceptability.

We don't know how successful this defense was, or even how strongly intuitionism could be said to have taken hold in Russia. By the end of the decade the main contributions by Russian mathematicians to constructivity were the results of Kolmogorov and Glivenko cited earlier in this book—both relatively minor theorems concerning Heyting's axiomatic system. And, the outright failure of Hilbert's program at the hands of Gödel would have negated the attempt to protect abstract mathematics from Marxist criticism. Indeed, mathematicians were not exempt from the Stalinist purges of the 1930s. In any event, intuitionism emerged in full in Russia in the latter half of the 1940s under the leadership of Andrey Markov, Jr., son of the renowned probability theorist Andrey Markov, Sr.

Markov, who had been attracted to intuitionistic mathematics already before the war, now decided to develop intuitionistic mathematics on a recursion-theoretic foundation using his kind of algorithms (now known as Markov algorithms). He embraced the identification of constructive functions with algorithmically computable ones through the adoption of a schematic version of the Church-Turing Thesis:

$$\forall x \forall y \big( A(x, y) \vee \neg A(x, y) \big) \wedge \forall x \exists y A(x, y) \rightarrow \exists f \, \forall x A(x, f(x)). \qquad \text{(CT)}$$

where the variables $x, y$ range over natural numbers and the variable $f$ ranges over functions computable by Markov algorithms, i.e., recursive (computable) functions. This acceptance would entail, like Brouwer's use of choice sequences, that Markov's constructivism would diverge from classical mathematics. Problems that are not solvable computably would allow provable instances of the rejection of PEM: $\neg\forall x\big(A(x) \vee \neg A(x)\big)$.

With this latter, one could also prove instances of the failure of the double negation rule: $\neg\forall x\big(\neg\neg\exists y A(x, y) \to \exists y A(x, y)\big)$. In fact, MP is a special case of this classical logical principle for decidable predicates.

The explicit acceptance of these two axioms CT and MP initially separates Markovian constructivity from intuitionism, which may choose to remain agnostic on these matters, even though Brouwer himself accepted the idea that we can construct sequences for which they do not hold. The treatment of real numbers leads to further separation. In classical mathematics, the real numbers can be obtained in a number of equivalent ways—via Cauchy sequences, Dedekind cuts, nested intervals, even decimal expansions. When intuitionistic reasoning is involved, the constructions can yield different structures. Indeed, as we saw in Sect. 4.5, Brouwer himself was the first to devise a weak counterexample to the claim that every real number has a decimal expansion. The construction of choice in intuitionistic mathematics is via Cauchy sequences, but even here there are differences and the Markov school has studied variants. Without getting into detail, let us say simply that the main variant consists of real numbers constructed as recursive sequences of rationals with recursive moduli of Cauchy convergence, studied constructively.

As in the West, the Russian introduction of the theory of algorithms took off in two directions—the development of recursion theory and the development of recursive analysis. Of interest here is the latter. Unlike Western recursive analysis, which was sporadic and unconnected (compare the work by Ernst Specker [77] and Reuben Goodstein [43]), the Russian development was systematic and consisted of a number of researchers working in various institutes in Moscow (led by Markov himself), Leningrad (led by Nikolai Shanin), Yerevan, Armenia (led by Igor Zaslavsky), and Prague, Czechoslovakia (led by Oswald Demuth), the height of its activities being the decades of the 1960 and 1970s.

The systematic nature of the Russian approach dictated that alternative mathematical constructions of the real numbers be considered. In addition to the basic construction of recursive Cauchy sequences of rational numbers with recursive moduli of convergence, they also considered, for example, the choice of recursive decimal expansions as constructive real numbers. This second continuum differs from the basic one in a number of ways. Recursion theory proves that the conversion from Cauchy sequence to decimal expansion is not algorithmic in nature and strong counterexamples are obtainable in Markovian recursive analysis. Indeed, we cannot resist mentioning the following cute result of Andrzej Mostowski and Yakov Uspensky (see [51, 52]):

**Theorem 6.27** (Mostowski-Uspensky) *Let $m, n > 1$ be integers. There is an algorithm translating every base-m expansion of a real number into a base-n expansion iff every prime divisor of $n$ is a divisor of $m$.*

In particular, the decimal expansions of real numbers are not algorithmically determinable. By CT this means that one can turn the weak counterexample of Brouwer into a strong one, a definite theorem: It is false that, for all real numbers $r$, $r$ has a decimal expansion.

In Markovian analysis we have Markov's Principle, which causes differences from Brouwerian analysis. We use the functional form of MP in its most pressing application, stated in the present book's Brouwerian fashion: Inequality and apartness relations for real numbers coincide, i.e.,

**Theorem 6.28**  $\text{MP} \leftrightarrow \forall x \in \mathbb{R}\ (x \neq 0 \leftrightarrow x \mathbin{\#} 0)$.

**Proof**  $\rightarrow$: Assume MP, and let $a \in \mathbb{R}$ be given by a sequence $(a_n)$. As we have seen before (Exercise 4.58), we can find a subsequence that converges at a prescribed rate. Assume that $\forall k \forall m > k(|a_k - a_m| < 2^{-k-1})$. We easily see that then $a = 0 \leftrightarrow \forall n(|a_n| < 2^{-n})$. Then $a \neq 0$ comes down to $\neg \forall n(|a_n| < 2^{-n})$, and so to $\neg\neg \exists n(|a_n| \geq 2^{-n})$. Apply MP: $\exists n(|a_n| \geq 2^{-n})$. From this immediately follows $\exists n(|a_n| > 2^{-n})$. Combined with $\forall k \forall m > k(|a_k - a_m| < 2^{-k-1})$ this yields $\exists p \forall q > p(|a_q| > 2^{-n-1})$, that is, $a \mathbin{\#} 0$. In the other direction, $a \mathbin{\#} 0 \rightarrow a \neq 0$ holds by the definition of #.

$\leftarrow$: Assume that $\forall x \in \mathbb{R}\ (x \neq 0 \leftrightarrow x \mathbin{\#} 0)$, and that $\neg\neg \exists x(\alpha(x) = 0)$. We define a Cauchy sequence as follows:

$$a_n = \begin{cases} 0 & \text{if } \forall m \leq n\ (\alpha(m) > 0) \\ 2^{-m} & \text{if } m < n \wedge \alpha(m) = 0 \wedge \forall p < m\ (\alpha(p) > 0). \end{cases}$$

This Cauchy sequence determines the number $a$.

$a = 0 \leftrightarrow \forall x(\alpha(x) > 0) \leftrightarrow \forall x \neg \alpha(x) = 0 \leftrightarrow \neg \exists x(\alpha(x) = 0)$.

So $\neg\neg \exists x(\alpha(x) = 0) \leftrightarrow a \neq 0 \leftrightarrow a \mathbin{\#} 0 \leftrightarrow \exists k(|a| > 2^{-k})$. It follows from the latter that a positive $\alpha(p)$ has already occurred before $\alpha(k)$, i.e., $\exists x(\alpha(x) = 0)$. $\qquad\square$

Perhaps this equivalence theorem leads someone to say that in Brouwer's mathematics MP can be refuted, in view of Theorem 6.18, which states that $\neg \forall x \in \mathbb{R}\ (x \neq 0 \leftrightarrow x \mathbin{\#} 0)$. Another argument would be that BKS is intuitionistically justified, but MP and BKS together imply PEM. For let $A$ be arbitrary. Apply BKS not to $A$, but to $A \vee \neg A$; this yields a BK-sequence $\alpha$ with $(A \vee \neg A) \leftrightarrow \exists n(\alpha(n) = 1)$. As $\neg\neg(A \vee \neg A)$ is a theorem, we have $\neg\neg \exists n(\alpha(n) = 1)$. Now, by MP, $\exists n(\alpha(n) = 1)$, whence $A \vee \neg A$.

However, these are only formal arguments, which do not take into account that Brouwer and Markov use their formulas to talk about different things. Brouwer's notion of infinite sequences is richer than what Markov was willing to accept. His $\mathbb{R}$ is therefore not the same as Markov's $\mathbb{R}$, and also the justification of BKS (from CS, the only known justification) depends on this richer notion. Conversely, MP was not formulated for application in a wider context than the recursive one to begin with. In his entry "Constructive trend" for the *Great Soviet Encyclopedia* (edition of 1979), Markov writes:

> The principal difference between constructivists and intuitionists is that constructivists, unlike intuitionists, do not think of their constructions as a purely mental pursuit. Moreover, intuitionists argue about certain "freely formed sequences" and consider the continuum as a "medium of free formation" and thus involve nonconstructive objects.

We conclude this section by remarking that with MP we see the same phenomenon we encountered with the full choice axiom AC in Sect. 6.8: in constructive settings, mathematics can influence logic. For further discussion of MP in relation to Brouwerian intuitionism, see [88] and the references there. For more on Russian constructivity, there are the books [51, 74]. Finally, MP is also important for the metamathematics of formal systems; but that is a topic outside the scope of this book.

### 6.9.2  Bishop's School

In the 1960s Errett Bishop formulated his own program for constructive mathematics, resulting in the book [8]. Bishop occupies an intermediate position; he wanted a notion of construction that would fit in with classical mathematics as well as with other constructive schools. He arrived at this by limiting the scope of classical concepts and theorems such that the result is evidently constructive to all, restricting for example, functions to continuous functions (and uniformly so in each closed interval, at that—no need to prove Brouwer's theorems, then). Thus, no conflicts with classical mathematics will arise, and at the same time there is plenty of room for weak counterexamples showing that some principles accepted in classical mathematics are not constructive.

It is clear that on this approach Brouwer's choice sequences and the intuitionistic analysis based on them are excluded. But Bishop did not insist on recursiveness either, taking the notions of construction and method to be the more basic ones. The mathematics he envisaged is often thought of as the part that classical mathematics, recursive mathematics, and intuitionistic mathematics have in common.[16]

Unfortunately, there is no Fan Theorem (Sect. 6.4).[17] This is not just because that theorem is false in recursive mathematics. While it is a theorem in both classical and intuitionistic mathematics, the respective proofs are very different, and both unacceptable in Bishop's setting: Brouwer's proof via his unorthodox proof of the

---

[16] The idea that the mathematics Bishop presented is the common part between classical and the various forms of constructive mathematics presents some challenges when going into detail (see [110, Sect. 8] and [67, "Bishop's constructive mathematics"]).

[17] Waaldijk [110] has observed that the definitions of continuity in [8, 10] are not the same, and has shown that proving their equivalence requires precisely the Fan Theorem.

Bar Theorem is not compatible with classical mathematics, and the classical proof is not constructive.

Bishop's ambitions and strength lay in his ability to introduce constructive methods into common modern mathematics, which "by default" was classical. He convincingly demonstrated that in analysis a constructive approach leads to strong, interesting and elegant theories and results. This also shows clearly in the work of mathematicians working in Bishop's style, such as Bridges and Richman.

On the other hand, Bishop did not invest much time in the analysis of the concept of constructivity, and steered clear of the philosophical foundational questions. There is a certain tendency in Bishop and others to see a contrast between giving attention to such matters and giving attention to "the mathematics itself" or "mathematical practice" or "the needs of mathematics". However, what counts as mathematics and the mathematical is itself a philosophical question.

Various theories have been developed that, either by intention or in effect, could count as formalizations of Bishop's mathematics. In fact, even though formalization is against the spirit of his book [8], Bishop himself made steps in that direction, for example in [9]. Of the others, we here mention Myhill's Constructive Set Theory [66], Feferman's Explicit Mathematics [37], and Martin-Löf's Type Theory [58]. In particular the latter project heavily reinvests in philosophical foundations, albeit of a rather different kind than Brouwer's.

## 6.10  Closing Remark

Markov's Recursive Mathematics and (formalizations of) Bishop's mathematics lend themselves very well to computational mathematics, favored by many contemporary constructivists. And understandably so. But for the field of constructivism as a whole, it would be an intellectual impoverishment if for that reason the more abstract or free parts of constructivism, the way to which was opened by Brouwer, would be left unexplored. Also, when thus going beyond computational mathematics, we should not, in turn, limit ourselves to seeing intuitionism's "potential" in terms of classical mathematics' "actual". It is, and has, a subject of its own.

# Appendix A
# Two Technical Devices in the Study of Intuitionistic Logic

**Abstract** A brief introduction to Gentzen's natural deduction and Kripke models.

It is a fact that, depending on one's interests and purposes, constructive logic has been much studied from different perspectives than Brouwer's. We here present two of the technical devices that have played a decisive role in that development: Gentzen's system of "natural deduction" for making formal derivations, and Kripke models in formal semantics (Kripke semantics).

## Gentzen's Natural Deduction

We have already seen that Brouwer's ideas about logic culminated in Heyting's Proof Interpretation. Now, despite all Brouwer's admonitions concerning language-less constructions, it is desirable to have a reasonably useful formal system of logical reasoning that is constructively acceptable. Such a system was presented by Gerhard Gentzen in 1934. It assumes that making inferences is the "core business" of logic, and not a nice piece of axiomatic practice, as in geometry for example. Gentzen's system of *Natural deduction* contained no axioms at all and only inference rules. The great clarity and the elegance of natural deduction lies, among other things, in the idea that deductions should only deal with only one inference at a time. How can one obtain a formula with a certain connective, and what do you know once you have such a formula? Gentzen's system is indeed "natural" because it mimics a way of thinking that is very natural to us. In particular, it allows us to reason not only from (added) axioms, but also from hypotheses. We are not going to give a course here, but we will provide sufficient material to give an insight into how natural deduction works.



D. van Dalen et al., *Intuitionistic Analysis*, Springer Undergraduate Mathematics Series, https://doi.org/10.1007/978-3-032-16491-9

Here is an example.

$$\frac{A \quad B}{A \wedge B} \wedge I \qquad \frac{A \wedge B}{A} \wedge E \qquad \frac{A \wedge B}{B} \wedge E$$

The notation explains itself: the formula below the line results from the formula(s) above it. The first inference rule introduces the connective for *and*, and the following rules eliminate the connective, according to the two possibilities. The rule on the left is called an *introduction rule*, the two rules on the right are called *elimination rules*.

The constructive interpretation of the above inference rules is "if there are constructions that make the formulas above the rule come out true, then there is a construction that makes the formula in the conclusion true". In the case of the conjunction this goes without saying: if there are proofs (constructions) $a : A$ and $b : B$, then there is a proof for $A \wedge B$.

Here are the natural deduction rules for propositional logic. The name is indicated after each rule. This is often useful for reading derivations.

$$\frac{A \quad B}{A \wedge B} \wedge I \qquad \frac{A \wedge B}{A} \wedge E \qquad \frac{A \wedge B}{B} \wedge E$$

$$\frac{A}{A \vee B} \vee I \qquad \frac{B}{A \vee B} \vee I \qquad \frac{A \vee B \quad \overset{[A]}{\underset{\vdots}{C}} \quad \overset{[B]}{\underset{\vdots}{C}}}{C} \vee E$$

$$\frac{\overset{[A]}{\underset{\vdots}{B}}}{A \to B} \to I \qquad \frac{A \quad A \to B}{B} \to E \qquad \frac{\bot}{A} \bot$$

When constructing proofs, it is convenient also to have

$$\frac{A \to B \quad A}{B} \to E$$

Logically speaking in this system the order of the two premises does not matter, but our writing system obliges us to choose one.

Recall (Sect. 2.2) that negation $\neg$ is defined as a special implication: $\neg A := A \to \bot$. Write down the resulting rules $\neg I$, $\neg E$ yourself.

The implication introduction rule requires some explanation; here we have something complicated above the line. The reading of the $\to I$ rule is as follows: if we have obtained a derivation of $B$ with the help of the permitted inference rules, assuming the hypothesis $A$, then we may conclude that $A \to B$ may be derived, whereby afterward the hypothesis $A$ is no longer needed. We say that among the hypotheses of the derivation, the hypothesis $A$ is canceled. In fact, we may keep the hypothesis $A$, but it is no longer needed for $A \to B$. This cancellation is indicated by enclosing the formula in question in square brackets.

The $\vee E$ rule even has two proofs above the line. There is an explanation for this: when we have a proof for $A \vee B$, that proof indicates one of the two disjuncts and adds a proof to it. If we now know in advance that $C$ follows from $A$ and other data, but that $C$ also follows from $B$ and other data, then we can at least derive $C$ no matter how a proof of $A \vee B$ is obtained.

The falsum rule says no more than that from the false, anything can be deduced.

A rule that is used in classical logic is that of *Reductio Ad Absurdum* (*RAA*):

$$\frac{\begin{array}{c} [\neg A] \\ \vdots \\ \bot \end{array}}{A}\ RAA$$

That formalizes proof by contradiction: if we can derive a contradiction from $\neg A$, then $A$ holds. We have seen that the proof by contradiction is not intuitionistically correct, and we therefore do not accept it. By the way, adding *RAA* to the usual formalization of intuitionistic logic[1] is enough to get the whole classical logic.

To illustrate, we present a couple of derivations. Which rules are used depends on the final formula (the formula to be proved). Once hypotheses have been suggested, they can be used again to work towards the conclusion, and so on.

We attach to each hypothesis an index, and the same one to the step where we cancel that hypothesis.

$$\frac{\dfrac{[A \wedge B]^1}{B}\wedge E \quad \dfrac{[A \wedge B]^1}{A}\wedge E}{\dfrac{B \wedge A}{A \wedge B \to B \wedge A}\to I_1}\wedge I \qquad \frac{\dfrac{\dfrac{[A]^2 \quad [A \to \bot]^1}{\bot}\to E}{(A \to \bot) \to \bot}\to I_1}{A \to ((A \to \bot) \to \bot)}\to I_2$$

That second derivation represents in effect our old argument to prove $A \to \neg\neg A$ (Sect. 2.2).

And here is a derivation for Brouwer's little theorem of 1908 that we saw in the same section. Now we use the abbreviation $\neg$ in the proof itself.

$$\frac{\dfrac{[\neg(A \vee \neg A)] \quad \dfrac{[A]}{A \vee \neg A}}{\dfrac{\bot}{\neg A}} \qquad \dfrac{[\neg(A \vee \neg A)] \quad \dfrac{[\neg A]}{A \vee \neg A}}{\dfrac{\bot}{\neg\neg A}}}{\dfrac{\bot}{\neg\neg(A \vee \neg A)}}$$

**Exercise A.1** In the above derivation, indicate at each step which rule has been used, and find the correct index for each hypothesis and the step where it is canceled. Write out the proof with $\to$ instead of $\neg$.

---

[1] However, for intuitionists logic is, like mathematics, in principle open-ended. So the usual formalization cannot be taken to be the definitive one. Heyting, who himself presented a formalism for intuitionistic logic, arithmetic, and a small part of analysis, warned about this from the beginning.

For the sake of completeness, we also add the derivation rules for the quantifiers.

$$\frac{A(x)}{\forall x\, A(x)}\ \forall I \qquad \frac{\forall x\, A(x)}{A(t)}\ \forall E$$

$$\frac{A(t)}{\exists x\, A(x)}\ \exists I \qquad \frac{\exists x\, A(x) \qquad \begin{array}{c}[A(x)]\\ \vdots\\ C\end{array}}{C}\ \exists E$$

With the quantifier rules we have to be careful with the variables. We demonstrate this with an application of $\forall I$ that goes wrong (again, indicate which rules have been used yourself):

$$\frac{\dfrac{\dfrac{[x=0]}{\forall x\,(x=0)}}{\dfrac{x=0 \rightarrow \forall x\,(x=0)}{\dfrac{\forall x\,(x=0 \rightarrow \forall x\,(x=0))}{0=0 \rightarrow \forall x\,(x=0)}}}}{}$$

The idea of the $\forall I$ rule is that $\forall x\, A(x)$ may be concluded from the preceding $A(x)$ because the $x$ does not have to satisfy any constraints. But here, in the top application of $\forall I$, $x$ is not "free" at all, for it is constrained to be equal to 0 ($x = 0$ here is an assumption!). The rest of the derivation is just there to show what strange conclusion we come to. And that happens only because we considered $x$ to represent an "arbitrary" number, when it clearly didn't.

We define the prohibition rules for the variables in the quantifier rules as follows:

in $\forall I$ and $\exists E$, the variable $x$ may not occur freely in a hypothesis on which $A(x)$ depends (i.e., a hypothesis that has not yet been canceled at that moment).
In $\forall E$, and $\exists I$, $t$ must be free for $x$.[2]

Natural deduction is an eminent example of a "do-it-yourself" system. The formulas to be proved (whether or not in combination with hypotheses) generally suggest what steps need to be taken in the derivation. For example, consider $(A \rightarrow B) \rightarrow ((B \rightarrow C) \rightarrow (A \rightarrow C))$. Since $\rightarrow$ is the main connective, it is obvious to use $A \rightarrow B$ as a hypothesis for an application of $\rightarrow I$, where $(B \rightarrow C) \rightarrow (A \rightarrow C)$ is the conclusion. Now the hypothesis can be "unpacked" by also assuming $A$ as a hypothesis.

The new conclusion again suggests applying the $\rightarrow I$ rule, now to $(B \rightarrow C)$ and $(A \rightarrow C)$. The new hypothesis $(B \rightarrow C)$ can be combined with $B$ in an application of $\rightarrow E$. Now the $\rightarrow I$ rule can be applied to $A$ and $C$, the result is $A \rightarrow C$. These intermediate steps can be found in the derivation below.

---

[2] "Free for" is a slightly technical term. Roughly speaking, it means that variables in $t$ should not get bound by quantifiers when substituted. See [102, p. 62].

$$\cfrac{\cfrac{\cfrac{[A]^1 \qquad [A \to B]^3}{B} \to E \qquad [B \to C]^2}{\cfrac{\cfrac{C}{A \to C} \to I_1}{(B \to C) \to (A \to C)} \to I_2}}{(A \to B) \to ((B \to C) \to (A \to C))} \to I_3 \to E$$

Note that by substituting $\bot$ for $C$ we get the derivation of $(A \to B) \to (\neg B \to \neg A)$.

Following Brouwer's idea "proofs are constructions", one can also use natural deduction to build up proof-constructions systematically. First an easy example:

$$\mathcal{D}_1 = A \qquad \mathcal{D}_2 = \frac{[A]}{A \to A}$$

To see that this is a correct derivation we must remember that above the line should be a derivation. Well, there is the derivation $\mathcal{D}_1$ where $A$ is both hypothesis and conclusion. It is the shortest derivation possible. When canceling, the hypothesis of $\mathcal{D}_1$ disappears and in the new derivation $\mathcal{D}_2$ becomes the premise of the implication, and the conclusion of $\mathcal{D}_1$ becomes the conclusion of the implication in $\mathcal{D}_2$.

Now we are going to assemble the corresponding proof-construction. We start at $\mathcal{D}_1$; there is no construction there yet because we have accepted $A$. This corresponds to a hypothetical construction $x$ of $A$, notation $x : A$. The proof-construction of $A \to A$ is a construction that transfers a hypothetical proof of the hypothesis (in our case $x$) into a proof of the conclusion of $\mathcal{D}_1$, that is again $x$. The identity $I$ therefore is a construction that does what is asked for. (There are other possibilities.) We make a schema as in the derivation $\mathcal{D}_2$ above:

$$\mathcal{D}_2 = \frac{[x : A]}{I : A \to A}$$

Now consider a general case of $\to I$:

$$\mathcal{D}_1 = \begin{array}{c} x : A \\ \vdots \\ t(x) : B \end{array} \qquad \mathcal{D}_2 = \frac{\begin{array}{c} [x : A] \\ \vdots \\ t(x) : B \end{array}}{\lambda x.t(x) : A \to B}$$

Explanation: in $\mathcal{D}_1$ the hypothesis $A$ is hypothetically proven by $x$, the conclusion by a construction built in the course of the derivation, and which we denote by the term $t(x)$. The proof construction of $A \to B$ is now the construction that transfers every possible proof $a$ into the proof $t(a)$.

To indicate such a proof we use $\lambda$-abstraction, which we encountered in Sect. 4.2 as a means of turning methods into functions, or more formally, terms into functions. Here, $t(x)$ is a term in which $x$ occurs as a free variable. Then $\lambda x.t(x)$ is the corresponding function, and we have $(\lambda x.t(x))(a) = t(a)$.

In the same way, we can make the operations on proof constructions involved in the other connectives explicit:

$$\frac{t_0 : A \qquad t_1 : B}{p(t_0, t_1) : A \wedge B} \wedge I \qquad \frac{t : A \wedge B}{p_0(t) : A} \wedge E \qquad \frac{t : A \wedge B}{p_1(t) : B} \wedge E$$

$$\frac{t(x^{[A]}) : B}{\lambda x.t(x) : A \to B} \to I \qquad \frac{s : A \qquad t : A \to B}{t(s) : B} \to E$$

$$\frac{t : A}{k_0(t) : A \vee B} \vee I \qquad \frac{t : B}{k_1(t) : A \vee B} \vee I$$

$$\frac{t : A \vee B \qquad t_0(x^A) : C \qquad t_1(y^B) : C}{D_{u,v}(t, t_0(u), t_1(v)) : C} \vee E$$

$$\frac{t : \bot}{\bot_A(t) : A} \bot$$

$$\frac{t(x) : A(x)}{\lambda y.t(y) : \forall x A(x)} \forall I \qquad \frac{t : \forall x A(x)}{t(s) : A(s)} \forall E$$

$$\frac{t : A(s)}{p(s, t) : \exists x A(x)} \exists I \qquad \frac{t : \exists x A(x) \qquad s(y, z^{A(y)}) : C}{E_{u,v}(t, s(u, v)) : C} \exists E$$

The rules now look slightly different from their pure natural deduction presentations. In particular, the sub-derivations have disappeared. That is because they are now hidden in the construction terms as variables.

In the rules for $\wedge$, we use the pair-forming and projection operators $p, p_0, p_1$ again that we saw in Sect. 4.2. In the rule for $\wedge$-introduction, $t_0$ and $t_1$ are variables for constructions that prove $A$ and $B$, respectively.

Also the elimination rules are in exact correspondence with the Proof Interpretation: if we have a construction for $A \wedge B$, then that is a pair, the first element of which is a proof of $A$.

The operators $k_0, k_1$ in the $\vee$-introduction serve to "obliterate case distinctions": if we have a proof of $A$, or we have a proof of $B$, then we also have a proof of $A \vee B$, regardless of the case we are in.

In $\vee$-elimination, the "invisible sub-derivations" are hidden in the variables $x$ and $y$. To indicate that they denote respective arbitrary constructions for $A$ and $B$, they have been given superscripts that contain that information. The construction in the conclusion contains a variable-binding operator $D$, which has the same function as the variable-binding lambda operator. The final construction for $C$ no longer depends on the hypothetical construction of $C$. The same consideration applies to $\exists$-elimination.

When applying the quantifier rules, one has to be a bit careful with the variables and terms. This also applies to the ordinary natural deduction, see e.g. [102, Sect. 3.8]. For the term calculus, as given above, see [82, Sect. 2.2].

Using the rules above, for each derivation the corresponding proof construction can be made. This is entirely in line with Brouwer's original idea that proofs are

**Fig. A.1** Stages 0 and 1 of the subject's activity

constructions. The idea of the proof terms is further elaborated in Per Martin-Löf's type theory [57, 58], in the typed lambda calculus, and other systems. The first to design and implement such a program was, in the late 1960s, Nicolaas de Bruijn, the father of the Automath project; see [109]. The so-called Curry-Howard isomorphism is also closely related to the Proof Interpretation, see [76].

## Kripke Models

Evert Beth and then Saul Kripke designed mathematical interpretations of Brouwer's logic. These forego essential questions—what are mental constructions, and how can we operate on them?—, yet to some extent model and summarize the conceptual ideas. They might in fact help to get a grip on the subject during a first acquaintance, and in any case they have been very fruitful for technical uses.

Both are based on a simplification of Brouwer's idea of the Creating Subject as an ideal mathematician or subject. We choose to focus on Kripke's semantics.

The subject's activity plays out in time—it can at any time direct its mathematical inquiry in any direction it wants. It can find out basic facts (by paying attention to what is given in intuition), and it can also begin to reason and arrive at more complicated conclusions. (We have seen a way of exploiting that idea within intuitionism itself in Sect. 6.5, in the form of the Brouwer-Kripke Schema.) We represent the subject's mathematical activity schematically in a diagram (Fig. A.1). The states are indicated with circles (black dots) and the continuation of the research in the future is indicated by a line. "Earlier" and "later" are indicated by "lower" and "higher" and a connecting line. Thus, we get a symbolic representation of the subject's activity. Much about this activity is a matter of choice, but not everything. For example, if at some point it determines that a number it has constructed is prime, it cannot discover at a later stage that it is divisible by 12.

We call the points in the diagram "nodes" or "states", and we denote them by the letters $k, \ell, m, \ldots$ with or without index. $k_0 < k_1$ means that the state $k_1$ comes after the state $k_0$. There is an idealizing assumption about the subject's thinking: it has a perfect memory, that is, it does not forget a fact once established. Technically stated, "truth is monotone". The notation for "the subject knows $A$ in state $k$" is $k \Vdash A$. This notation comes from set theory, and there is a good reason to keep it here. We will pronounce $k \Vdash A$ as "$k$ *makes A true*" or also "$k$ *forces A*".

**Fig. A.2**  Model 1

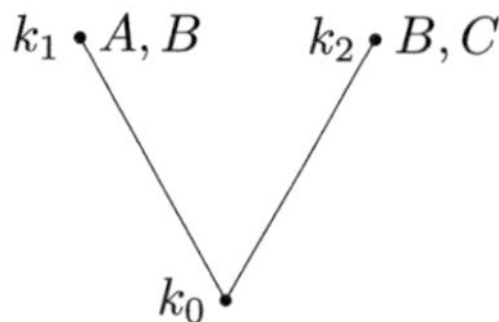


We must now try to see how the subject can come to know compound statements. For $\wedge$ this is not so difficult, the subject knows (or "has made evident") $A \wedge B$ at the moment $k$ if it knows both statements, in other words $k \Vdash A \wedge B$ as $k \Vdash A$ and $k \Vdash B$. For $\vee$ the situation is exactly like this: $k \Vdash A \vee B$ if $k \Vdash A$ or $k \Vdash B$. In words: you know $A \vee B$ at moment $k$ if you know at least one of them at that moment.

How to model implication? The answer is as natural as it is simple: in the state $k$, the subject knows that $A \to B$ when it determines that as soon as $A$ is known in a future situation $\ell$, $B$ is also known. So we define:

$$k \Vdash A \to B \text{ if for all } \ell \text{ with } k \leq \ell, \text{ if } \ell \Vdash A \text{ then } \ell \Vdash B$$

We now give a formal definition of $\Vdash$, i.e., the rendition in these models of the intuitionistic concept of truth, but using the logical signs in the metalanguage. We use $D(k)$ for the domain that the subject is in at a node $k$. Note that in general the domains can grow: $k \leq \ell \to D(k) \subseteq D(\ell))$.[3]

**Definition A.2**  *Truth at a node* is captured as follows:

$$k \Vdash A \wedge B := k \Vdash A \text{ and } k \Vdash B$$
$$k \Vdash A \vee B := k \Vdash A \text{ or } k \Vdash B$$
$$k \Vdash A \to B := \forall \ell \geq k \, (\ell \Vdash A \to \ell \Vdash B)$$
$$k \Vdash \neg A := \forall \, \ell \geq k \, \ell \nVdash A$$
$$k \Vdash \forall x A(x) := \forall \, \ell \geq k \, \forall a \in D(\ell) \, \ell \Vdash A(a)$$
$$k \Vdash \exists x A(x) := \exists A \in D(k) \, k \Vdash A(a)$$

and also

$$k \nVdash \bot .$$

Look at model 1 (Fig. A.2). At moment $k_0$ the subject knows nothing about $A, B, C$. There are two continuations of the investigation; in one, $k_1$, it finds $A$ and $B$, in the other, $k_2$, $B$ and $C$. According to our definition, $k_0 \Vdash A \to B$ now holds, but, for example, $k_0 \Vdash B \to A$ does not hold. Assuming that we have honestly written down what the subject learns, we can determine that for example $k_0 \Vdash \neg D$, since neither at $k_0$ nor at its successors, $k_1, k_2$, is $D$ known.

---

[3] Unlike Brouwerian intuitionism, Kripke semantics does not take the idea of growth literally. While it understands growth as an inclusion relation between certain sets, metaphysically these sets may exist outside of time altogether.

As in classical logic, we define the equivalence $A \leftrightarrow B$ as $(A \rightarrow B) \wedge (B \rightarrow A)$.

Let us take a closer look at the negation. The fact that the subject doesn't know $A$ at a given moment doesn't mean much; perhaps it finds evidence for $A$ moments later. Only when it knows that it will never find $A$ at any time does it know that $A$ is impossible, i.e., that $\neg A$ is true. Here we are again dealing with the future. The subject now knows $\neg A$ if it can see that $A$ will not hold at any later time. Hence, the above definition:

$$k \Vdash \neg A \text{ if for all } \ell \text{ with } \ell \geq k \; (\ell \not\Vdash A)$$

There is a connection between the implication and the negation. For convenience, we introduce a symbol for the prototype of contradiction: $\bot$. How you want to think about this contradiction is not so important. Most people think of something concrete, such as 0=1. But there are many other possibilities. The most important property of the contradiction is that it can never be true, in other words

$$k \Vdash \bot \text{ holds for no } k$$

**Exercise A.3** Prove:

(1) $k \Vdash A \leftrightarrow B \leftrightarrow \forall \ell \geq k \; (\ell \Vdash A \leftrightarrow \ell \Vdash B)$,
(2) $\forall k \;\; k \Vdash \neg A \leftrightarrow (A \rightarrow \bot)$.

To be clear: we always talk about forcing, or about truth in a situation (at a stage, at a moment), without specifying how many situations there are, and how they relate in terms of "earlier" and "later". That is not always necessary. Just as you can talk about arbitrary functions, triangles, etc., you can also talk about arbitrary collections of states and time divisions. The official notion is that of *Kripke model*.

**Definition A.4** A *Kripke model* is a set with a partial ordering $\leq$,[4] together with a function $f$ that adds to each node $k$ a set of atomic statements $A$. It is assumed that $f(k) \subseteq f(\ell)$ whenever $k \leq \ell$. We write $k \Vdash A$ if $A$ is in the set added to $k$. Thus, if $k \leq \ell$ and $k \Vdash A$, then $\ell \Vdash A$. When dealing with predicate logic, furthermore a set $D(k)$, the domain of $k$, is added to each node $k$. For these domains we have $k \leq \ell \rightarrow D(k) \subseteq D(\ell)$.

We have already shown above how the Kripke model is a rough way of indicating the mathematical activity of the subject; forcing played the role of "acquisition of knowledge". The subject also creates objects in the process; they are not lost at later times. These elements will not play a role in the interpretation of propositional logic, but they do when we look at predicate logic below.

This definition may be a little daunting, but it is simply a general definition of what we called diagrams above.

---

[4] A partial ordering $\leq$ satisfies transitivity $(k \leq \ell \wedge \ell \leq m \rightarrow k \leq m)$ and antisymmetry $(k \leq \ell \wedge \ell \leq k \rightarrow k = \ell)$. Think of "less than or equal to" for numbers, or "is divisible by" for positive numbers, or "subset of" for sets.

**Lemma A.5** *With a given Kripke model, the forcing relation can be extended to all formulas based on the Definition A.2.*

The proof of the lemma is trivial, as the definition was designed with that in mind.

Now let us go back to model 1 that we gave earlier in this section. We can already test many formulas for their truth. E.g., $k_1 \Vdash \neg C$, because neither in $k_1$ nor in a later $\ell$ (there is none) is $C$ is forced.

**Exercise A.6**  Check in which nodes of model 1 the following formulas are forced.

(1)  $B \to A$,
(2)  $\neg A$,
(3)  $A \vee B$,
(4)  $\neg\neg B$,
(5)  $\neg A \to B$,
(6)  $A \vee \neg A$,
(7)  $\neg\neg B \to B$,
(8)  $\neg(A \wedge C)$,
(9)  $\neg A \vee \neg B$,
(10)  $\neg(A \wedge C) \to (\neg A \vee \neg B)$,
(11)  $\neg(\neg A \wedge \neg B)$.

The next lemma is less trivial; it says that the forcing relation is monotone for all formulas.

**Lemma A.7**  *Let A be a formula not necessarily atomic. If $k \Vdash A$ and $k \leq \ell$ holds in a Kripke model, then also $\ell \Vdash A$.*

***Proof***  We show two cases.

($\to$)  Given $k \Vdash A \to B$ and $k \leq k_1$. To show that $k_1$ also forces $A \to B$, we have to show that in every $k_2 \geq k_1$ from $k_2 \Vdash A$ follows $k_2 \Vdash B$. Well, if $k_2 \geq k_1$, then also $k_2 \geq k$. Now we use that $k \Vdash A \to B$; by definition $k_2 \Vdash B$ holds if $k_2 \Vdash A$ does.

($\forall$)  Given $k \Vdash \forall x A(x)$ and $k \leq k_1$. We have to show that for every $k_2 \geq k_1$ and for every $a \in D(k_2)$ we have $k_2 \Vdash A(a)$. Well, we know that $k_2 \geq k$, and so by definition $k_2 \Vdash A(a)$ holds.

The other cases are simpler because there the future plays no role.  $\square$

When can we say that a statement $A$ is true? This is obvious: if in every investigation by the subject $A$ is true under every condition. More precisely:

**Definition A.8**  *$A$ is true if, for each node $k$ in any Kripke model, $k \Vdash A$.*

Perhaps this definition seems hopeless: how can we know what is happening in all Kripke models? However, the situation is, in fact, not bad. We have been used to general reasoning for centuries. "All people have hearts" is experienced as true, although we have not cut all people open to carry out the check. It is rather because we know something about the nature or essence or definition of humans. In the same manner we can reason about all Kripke models without inspecting them one by one. Just consider the following statement:

$$A \wedge B \rightarrow A \text{ is true.} \quad (1)$$

How could we prove that claim? Well, it is very simple.

***Proof*** Suppose you have a Kripke model, and that for a node $\ell$ in it, $\ell \Vdash A \wedge B$. By definition, that means $\ell \Vdash A$ and $\ell \Vdash B$. But then we're done, because then we at least know $\ell \Vdash A$. Now the definition of forcing for the implication says take a node $k$ in the model, if for every $\ell \geq k$ we have $\ell \Vdash A$ if $\ell \Vdash A \wedge B$, then $k \Vdash A \wedge B \rightarrow A$ holds. But we just established that. $\qquad\qquad\square$

Another example:
$$A \rightarrow \neg\neg A \text{ is true.} \quad (2)$$

***Proof*** Let us first define a useful criterion for forcing $\neg\neg A$: $k \Vdash \neg\neg A$ if and only if for all $\ell \geq k$ $\ell \nVdash \neg A$. Now $\ell \nVdash \neg A$ means that there is an $m \geq \ell$ with $m \Vdash A$. We allow ourselves classical logic here! But that is only natural, in so far as Kripke semantics are, as we indicated in the introduction to this section and in footnote 3, essentially a classical approximation of intuitionistic ideas. So,

$$k \Vdash \neg\neg A \leftrightarrow \forall \ell \geq k \; \exists m \geq \ell \; (m \Vdash A)$$

Now the truth of $k \Vdash A \rightarrow \neg\neg A$ can be seen almost immediately. After all we want that above every node above $k$ there is a node where $A$ is forced. But that is obvious because $A$ is forced into every node above $k$.[5] $\qquad\qquad\square$

**Exercise A.9** Construct a Kripke model with two nodes in which $\neg\neg A$ but not $A$ is forced.

Finally, we show the truth of one of De Morgan's two laws.

$$\neg(A \vee B) \leftrightarrow (\neg A \wedge \neg B).$$

We first consider the implication from left to right, and assume that in a Kripke model $k \Vdash \neg(A \vee B)$ holds. The proof is going to be indirect. Suppose that $k \nVdash (\neg A \wedge \neg B)$, then $k$ should not force one of the two conjuncts. Say $k \nVdash \neg B$, then there is an $\ell$ above $k$ with $\ell \Vdash B$. By definition $\ell \Vdash A \vee B$ now also holds, but this conflicts with $k \Vdash \neg(A \vee B)$. Had we chosen $A$ instead of $B$, we would have encountered the same contradiction. So the assumption $k \nVdash (\neg A \wedge \neg B)$ was incorrect, so $k \Vdash (\neg A \wedge \neg B)$.

---

[5] Note that, for the sake of convenient wording, we've used "above" here for "higher or equal".

**Fig. A.3** Model 2

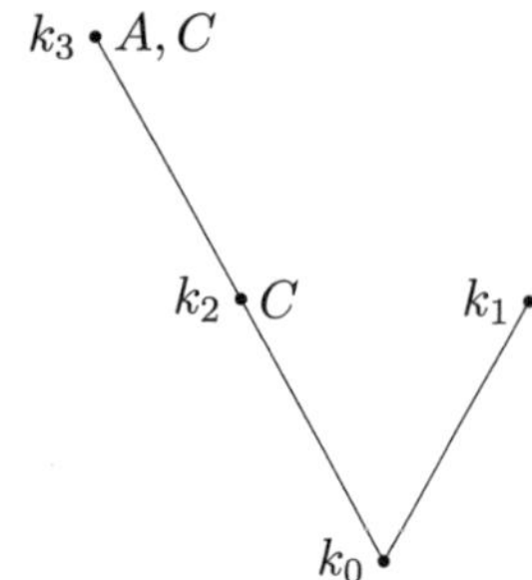


## Exercise A.10

1. A moment ago, we used RAA. Give a constructive argument instead.
2. Prove the implication from right to left yourself.

**Exercise A.11** Show that De Morgan's other law, $\neg(A \wedge B) \to (\neg A \vee \neg B)$, is not constructive.

To do this, create a Kripke model with three nodes âŁ" a smallest node with two incomparable nodes above it. Cleverly choose at which nodes $A$ and $B$ are true.

**Exercise A.12** Prove the truth of the following formulas.

(1) $(A \wedge B) \to (A \vee B)$
(2) $A \to (\neg A \to B)$
(3) $[A \wedge (A \to B)] \to B$

We list the most important formulas that are valid classically but not constructively.

(1) $k_0 \nVdash A \vee \neg A$
(2) $k_0 \nVdash \neg\neg B \to B$
(3) $k_0 \nVdash (A \to B) \to (\neg A \vee B)$
(4) $k_0 \nVdash \neg(\neg A \wedge \neg C) \to (A \vee C)$
(5) $k_0 \nVdash \neg A \vee \neg\neg A$
(6) $k_0 \nVdash (\neg A \to \neg C) \to (C \to A)$

With model 1 from above, we can show that (1) through (5) are not generally valid, for (6) we use model 2 (Fig. A.3).

The reader will soon see that some weak counterexamples (i.e., Kripke models that do not make a certain formula true) can be made with fewer nodes. A good exercise.

The Kripke models again demonstrate that from a constructive standpoint we have to regard ordinary logic with a fair dose of mistrust.

Note that the example (6) above shows that the "contraposition" only works one way. Intuitively, $(A \to B) \to (\neg B \to \neg A)$ is correct (a simple argument shows that if we take $A \to B$ and $\neg B$, a contradiction can be derived from $A$). Everyday counterexamples can usually be found, but they are not always equally natural. Here

**Fig. A.4** Model 3

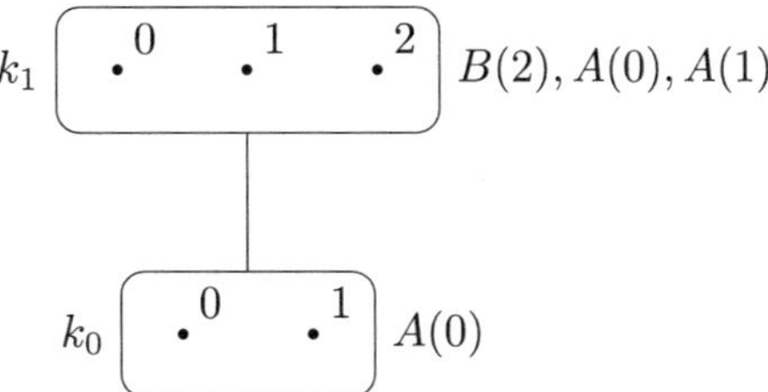


is one for the invalid direction of the contraposition: If from the fact that Jane could not solve the chess problem from last week it follows that she had not concentrated, then we cannot conclude that Jane could have found the solution if she had concentrated. Try to find other everyday examples yourself.

Coming to predicate logic, it now is important to take into consideration elements of each node. They represent the objects that the subject has constructed. We therefore now call the nodes *(possible) worlds*. Each node $k$ has a world $D(k)$ ("$D$" for domain). The elements can have properties $A(x), B(x), C(x), \ldots$. The convention is that (i) the subject does not forget anything it has once established: $k \leq \ell \wedge k \Vdash A \rightarrow \ell \Vdash A$, (ii) the subject does not destroy anything it once created: $k \leq \ell \wedge a \in D(k) \rightarrow a \in D(\ell)$. Popularly said, the worlds grow over time, or they at least don't get smaller.

"For all $x$" now should mean "for all elements the subject has now and will make in the future"; "there is an $x$" is a stronger statement, the claimed element must *now* already exist, and should not be dismissed with a promise for the future. Thus, we define

$$k \Vdash \forall x\, A(x) \leftrightarrow \forall \ell \geq k\ \forall a \in D(\ell)\ \ell \Vdash A(a)$$
$$k \Vdash \exists x\, A(x) \leftrightarrow \exists a \in D(k) \text{ such that } k \Vdash A(a)$$

Model 3 (Fig. A.4) will be used as an example. In it,

$$k_0 \Vdash \exists x\, A(x)\,,$$
$$k_0 \nVdash \exists x\, B(x)\,,$$
$$k_0 \Vdash \neg\neg\exists x\, B(x)\,,$$
$$k_0 \nVdash \exists x\, \neg A(x)\,,$$
$$k_0 \Vdash \neg\forall x\, A(x)\,.$$

So we see in particular that $k_0 \nVdash \neg\forall x\, A(x) \rightarrow \exists x\, \neg A(x)$.

Make your own model in which $\neg\forall x\, \neg A(x) \rightarrow \exists x\, A(x)$ does not hold; yet another classical law turns out to be unreliable.

**Exercise A.13**  Show that the following formulas are true:

(1)  $\forall x(A(x) \wedge B(x)) \leftrightarrow \forall x A(x) \wedge \forall y B(y)$
(2)  $\neg \exists x A(x) \leftrightarrow \forall x \neg A(x)$
(3)  $\exists x(A(x) \vee B(x)) \leftrightarrow \exists x A(x) \vee \exists y B(y)$
(4)  $\exists x A(x) \rightarrow \neg \forall x \neg A(x)$

**Exercise A.14**  Give Kripke models in which the following formulas are false:

(1)  $\neg \forall x \neg A(x) \rightarrow \exists x A(x)$
(2)  $\forall x(A(x) \vee B) \rightarrow \forall x A(x) \vee B$, where $x$ does not occur in $B$.
(3)  $\forall x \forall y(x = y \vee x \neq y)^6$
(4)  $(A \rightarrow \exists x B(x)) \rightarrow \exists x(A \rightarrow B(x))$

---

$^6$ The symbol $\neq$ stands for "unequal", or "not equal". Recall that the equality is an ordinary binary relation.

# References and Further Reading

Depending on the reader's background and interests, there are many papers and books that we would like to recommend for fruitful further reading. Here we must present a choice. It lies in the nature of Brouwer's enterprise that most literature on it has, to varying extents, both a mathematical and a philosophical character.

Easily accessible in several senses are the following entries on intuitionistic and constructive topics in two online encyclopedias. In the *Stanford Encyclopedia of Philosophy*: [6, 14, 35, 49, 63, 89, 90]. In the *Internet Encyclopedia of Philosophy*: [7, 61]. More technical material can be found at nLab, "a wiki for collaborative work on Mathematics, Physics, and Philosophy" [67].

In Edwards' *Encyclopedia of Philosophy*, Parsons' articles [70, 71] are highly recommended. (In the second edition, these have been augmented.)

Most of Brouwer's published papers can be found in [28, 29]; of various papers that are not in English, translations can be found in [55, 103].

Book-length presentations of intuitionism by Brouwer himself are his posthumous [30, 31]. Of book-length introductions to intuitionism by others, [47] is a classic. More recent are [34, 72, 84, 105]. More technical, yet also of philosophical interest are [33, 80, 81, 107]. A shorter overview is [64]. Extensive mathematical and philosophical treatments of Brouwer's "Creating Subject" are [59, 88].

Aiming at classical, formally inclined mathematicians is the rather technical but rich and influential [50]. While the authors stay away from Brouwer's foundational ideas properly speaking, their demonstration of the independence of the Bar Theorem and of the consistency of bar induction relative to a certain classical system did help to make it clear to classical mathematicians that there really was something to intuitionistic mathematics.[7] (They also included the continuity principle that after the proof of Theorem 6.19 we argued Brouwer would not have accepted.) The book also refers to more of Brouwer's and other intuitionists' work than people were, and are, in general familiar with.

---

[7] We thank an anonymous reader for emphasizing this to us.



D. van Dalen et al., *Intuitionistic Analysis*, Springer Undergraduate Mathematics Series, https://doi.org/10.1007/978-3-032-16491-9

153

Brouwer's life is described in detail in [96, 97]; second edition, in one volume: [101]. The reception of Brouwer's work in the 1920s is the topic of [45]. For historical and philosophical aspects of Gödel's interest in intuitionism, see the part on Brouwer in [87] and the references there.

Brouwer's philosophy of mathematics is treated from the perspective of Husserl's phenomenology in [85] and [87, Chap. 12], and from that of Eastern philosophy in [69].

Of general works on constructive mathematics, we mention [5, 15, 56, 62, 79, 83]. An enjoyable text written especially for the benefit of classical mathematicians is [4].

On the Bishop school: [10, 16]. General, but with emphasis on Bishop: [13]; Crosilla's chapter there is a philosophical discussion of Bishop's work. The Russian school: [51–53, 74].

For introductions to constructivism in category theory and type theory, we refer to [38, 41, 54, 57, 58, 68, 104].

A recent survey of the status of the Axiom of Choice in a variety of intuitionistic and constructive systems is [60].

For some recent ideas on intuitionistic mathematics in physics, see [39].

# References


1. Barendregt, H.: Kreisel, lambda calculus, a windmill and a castle. In: Odifreddi, P. (ed.) Kreiseliana. About and around Georg Kreisel, pp. 3–14. A K Peters, Wellesley, MA (1996)
2. Barzin, M., Errera, A.: Sur la logique de M. Brouwer. Académie Royale de Belgique, Bulletin de la classe des sciences **13**(5), 56–71 (1927)
3. Barzin, M., Errera, A.: Sur le principe du tiers exclu. Archives de la Société Belge de Philosophie **1**, 3–26 (1929). Fascicule 2
4. Bauer, A.: Five stages of accepting constructive mathematics. Bull. Am. Math. Soc. New Ser. **54**, 481–498 (2017)
5. Beeson, M.: Foundations of Constructive Mathematics. Springer, Heidelberg (1985)
6. Bell, J.: Continuity and infinitesimals. In: Zalta, E. (ed.) The Stanford Encyclopedia of Philosophy, Spring 2022 edn. Metaphysics Research Lab, Stanford University (2022). https://plato.stanford.edu/archives/spr2022/entries/continuity/
7. Bentzen, B.: Intuitionism in mathematics. In: Fieser, J., Dowden, B. (eds.) The Internet Encyclopedia of Philosophy. https://iep.utm.edu/intuitionism-math/
8. Bishop, E.: Foundations of Constructive Analysis. McGraw-Hill, New York (1967)
9. Bishop, E.: Mathematics as a numerical language. In: Kino, A., Myhill, J., Vesley, R. (eds.) Intuitionism and Proof Theory. Proceedings of the Summer Conference at Buffalo N. Y., 1968, pp. 53–71. North-Holland, Amsterdam (1970)
10. Bishop, E., Bridges, D.: Constructive Analysis. Springer, Heidelberg (1985)
11. Borchert, D. (ed.): Encyclopedia of Philosophy, 2nd edn. Thomson Gale / Macmillan Reference USA, Detroit (2006)
12. Borwein, J.: Brouwer-Heyting sequences converge. Mathematical Intelligencer **20**(1), 14–15 (1998)
13. Bridges, D., Ishihara, H., Rathjen, M., Schwichtenberg, H.: Handbook of Constructive Mathematics, Encyclopedia of Mathematics and Its Applications, vol. 185. Cambridge University Press, Cambridge (2023)

14. Bridges, D., Palmgren, E., Ishihara, H.: Constructive mathematics. In: Zalta, E., Nodelman, U. (eds.) The Stanford Encyclopedia of Philosophy, Fall 2022 edn. Metaphysics Research Lab, Stanford University (2022). https://plato.stanford.edu/archives/fall2022/entries/mathematics-constructive/

15. Bridges, D., Richman, F.: Varieties of Constructive Mathematics. No. 97 in London Math. Soc. Lecture Notes. Cambridge University Press (1987)

16. Bridges, D., Vîţă, L.: Techniques of Constructive Analysis. Springer, New York, NY (2006)

17. Brouwer, L.E.J.: Over de grondslagen der wiskunde. Ph.D. thesis, Universiteit van Amsterdam (1907). Translation "On the Foundations of Mathematics" in [28, pp. 11–101], with original page numbers in the margin

18. Brouwer, L.E.J.: De onbetrouwbaarheid der logische principes. Tijdschrift voor Wijsbegeerte **2**, 152–158 (1908). Two different translations: [28, pp. 107–111] and [93]

19. Brouwer, L.E.J.: Beweis der Invarianz der Dimensionenzahl. Mathematische Annalen **70**, 161–165 (1911). Also in [29, pp. 430–434]

20. Brouwer, L.E.J.: Über Abbildungen von Mannigfaltigkeiten. Mathematische Annalen **71**, 97–115 (1911). Also in [29, pp. 454–472]

21. Brouwer, L.E.J.: Begründung der Mengenlehre unabhängig vom logischen Satz vom ausgeschlossenen Dritten. Erster Teil: Allgemeine Mengenlehre. KNAW Verhandelingen **5**, 1–43 (1918). Also in [28, pp. 150–190]

22. Brouwer, L.E.J.: Besitzt jede reelle Zahl eine Dezimalbruchentwicklung? Mathematische Annalen **83**, 201–210 (1921). Also in [28, pp. 236–245]. Translation [55, pp. 28–35]

23. Brouwer, L.E.J.: Intuitionistische Zerlegung mathematischer Grundbegriffe. Jahresbericht der Deutschen Mathematiker Vereinigung **33**, 251–256 (1925). Also in [28, pp 275–280]. Translation [55, pp. 287–289 (sections 2–4), pp. 290–292 (section 1)]

24. Brouwer, L.E.J.: [Notes for the graduation speech to Wilfrid Wilson on February 10] (1928). Noord-Hollands Archief, Haarlem. Collection 889, folder 199. The archive put a scan online as page 40 of the document at https://hdl.handle.net/21.12102/544EA3FE00F04716AA2485A240CF110C

25. Brouwer, L.E.J.: Door klassieke theorema's gesignaleerde pinkernen die onvindbaar zijn. Indagationes Mathematicae **14**, 443–445 (1952). Translation in [28, pp. 519–521]

26. Brouwer, L.E.J.: An intuitionistic correction of the fixed-point theorem on the sphere. Proc. R. Soc. Lond. Ser. A **213**, 1–2 (1952). Also in [28, pp. 506–507]

27. Brouwer, L.E.J.: Points and spaces. Can. J. Math. **6**, 1–17 (1954). Also in [28, pp. 522–540]

28. Brouwer, L.E.J.: Philosophy and Foundations of Mathematics (ed. A. Heyting), Collected Works, vol. 1. North-Holland, Amsterdam (1975)

29. Brouwer, L.E.J.: Geometry, Analysis, Topology and Mechanics (ed. H. Freudenthal), Collected Works, vol. 2. North-Holland, Amsterdam (1976)

30. Brouwer, L.E.J.: Brouwer's Cambridge Lectures on Intuitionism (ed. D. van Dalen), Cambridge University Press, Cambridge (1981)

31. Brouwer, L.E.J.: Intuitionismus (ed. D. van Dalen), Bibliographisches Institut. Wissenschaftsverlag, Mannheim (1992)

32. Cohen, L., Forster, Y., Kirst, D., Paiva, B.D.R., Rahli, V.: Separating Markov's principles. In: LICS 2024: Proceedings of the 39th Annual ACM/IEEE Symposium on Logic in Computer Science, pp. 1–14. ACM, New York (2024)

33. Dragálin, A.: Mathematical Intuitionism. Introduction to Proof Theory. American Mathematical Society, Providence, RI (1988). Original publication Moscow, 1979

34. Dummett, M.: Elements of Intuitionism, 2nd, rev Clarendon Press, Oxford (2000)

35. Dybjer, P., Palmgren, E.: Intuitionistic type theory. In: Zalta, E. (ed.) The Stanford Encyclopedia of Philosophy, Spring 2023 edn. Metaphysics Research Lab, Stanford University (2023). https://plato.stanford.edu/archives/spr2023/entries/type-theory-intuitionistic/

36. Edwards, P. (ed.): The Encyclopedia of Philosophy. Macmillan, New York (1967)

37. Feferman, S.: A language and axioms for explicit mathematics. In: Crossley, J. (ed.) Algebra and Logic. Proceedings 1974. Springer Lecture Notes in Mathematics, vol. 450, pp. 87–139. Monash University, Australia (1975)

38. Girard, J.Y., Lafont, Y., Taylor, P.: Proofs and Types. Cambridge University Press, Cambridge (1989)

39. Gisin, N.: Indeterminism in physics and intuitionistic mathematics. Synthese **199**, 13345–13371 (2021)

40. Glivenko, V.: Sur quelques points de la logique de M. Brouwer. Académie Royale de Belgique, Bulletin de la classe des sciences **5**(15), 183–188 (1929). Translation [55, pp. 301–305]

41. Goldblatt, R.: Topoi. The Categorial Analysis of Logic. North-Holland, Amsterdam (1979). Revised 1983. Dover edition 2006

42. Goodman, N., Myhill, J.: Choice implies excluded middle. Zeitschrift für mathematische Logik und Grundlagen der Mathematik **24**(5), 461 (1978)

43. Goodstein, R.: Recursive Analysis. Elsevier, Amsterdam (1961)

44. Heinzmann, G.: Poincaré, Russell, Zermelo et Peano. Textes de la discussion (1906-1912) sur les fondements des mathématiques: des antinomies à la prédicativité. Blanchard, Paris (1986)

45. Hesseling, D.: Gnomes in the Fog. The Reception of Brouwer's Intuitionism in the 1920s. Birkhäuser, Basel (2003)

46. Heyting, A.: Bemerkungen zu dem Aufsatz von Herrn Freudenthal "Zur intuitionistischen Deutung logischer Formeln". Compos. Math. **4**, 117–118 (1937)

47. Heyting, A.: Intuitionism. An Introduction. North-Holland, Amsterdam (1956)

48. Heyting, A.: Intuitionism. An Introduction, 3rd rev. edn. North-Holland, Amsterdam (1971)

49. Iemhoff, R.: Intuitionism in the philosophy of mathematics. In: Zalta, E. (ed.) The Stanford Encyclopedia of Philosophy, Fall 2020 edn. Metaphysics Research Lab, Stanford University (2020). https://plato.stanford.edu/archives/fall2020/entries/intuitionism/

50. Kleene, S., Vesley, R.: The Foundations of Intuitionistic Mathematics. Especially in Relation to Recursive Functions. North-Holland, Amsterdam (1965)

51. Kushner, B.: Lectures on Constructive Mathematical Analysis. American Mathematical Society, Providence, RI (1984). Translation of Russian edition, 1973

52. Kushner, B.: Markov's constructive analysis; a participant's view. Theor. Comput. Sci. **219**(1–2), 267–285 (1999)

53. Kushner, B.: The constructive mathematics of A. A. Markov. Am. Math. Mon. **113**(6), 559–566 (2006)

54. Mac Lane, S., Moerdijk, I.: Sheaves in Geometry and Logic. A First Introduction to Topos Theory. Springer, New York (1992)

55. Mancosu, P.: From Brouwer to Hilbert. The Debate on the Foundations of Mathematics in the 1920s. Oxford University Press, Oxford (1998)

56. Martin-Löf, P.: Notes on Constructive Mathematics. Almqvist & Wiksell (1970)

57. Martin-Löf, P.: Intuitionistic Type Theory. Notes by G. Sambin of a Series of Lectures Given in Padua, June 1980. Bibliopolis, Napoli (1984)

58. Martin-Löf, P.: An intuitionistic theory of types: predicative part. In: Rose, H., Shepherdson, J. (eds.) Logic Colloquium '73, pp. 73–118. North-Holland, Amsterdam (1975)

59. Martino, E.: Intuitionistic Proof versus Classical Truth. The Role of Brouwer's Creative Subject in Intuitionistic Mathematics. Springer, Cham (2018)

60. McCarty, C., Shapiro, S., Klev, A.: The axiom of choice is false intuitionistically (in most contexts). Bulletin of Symbolic Logic **29**(1), 71–96 (2023)

61. McKubre-Jordens, M.: Constructive mathematics. In: Fieser, J., Dowden, B. (eds.) The Internet Encyclopedia of Philosophy. https://iep.utm.edu/constructive-mathematics/

62. Mines, R., Richman, F., Ruitenburg, W.: A Course in Constructive Algebra. Springer, New York (1988)

63. Moschovakis, J.: Intuitionistic logic. In: Zalta, E. (ed.) The Stanford Encyclopedia of Philosophy, Fall 2021 edn. Metaphysics Research Lab, Stanford University (2021). https://plato.stanford.edu/archives/fall2021/entries/logic-intuitionistic/

64. Moschovakis, J., Vafeiadou, G.: Intuitionistic Mathematics and Logic (2020). https://arxiv.org/abs/2003.01935

65. Myhill, J.: Notes towards an axiomatization of intuitionistic analysis. Logique et Anal. (N.S.) **35**, 280–297 (1966)

66. Myhill, J.: Constructive set theory. J. Symb. Log. **40**, 347–382 (1975)
67. nLab authors: nLab. https://ncatlab.org/nlab/show/HomePage
68. Nordström, B., Petersson, K., Smith, J.: Programming in Martin-Löf's Type Theory. An Introduction. Oxford University Press, Oxford (1990)
69. Pambuccian, V.: Brouwer's Intuitionism: Mathematics in the being mode of existence. In: Sriraman, B. (ed.) Handbook of the History and Philosophy of Mathematical Practice, pp. 645–699. Springer, Cham (2024)
70. Parsons, C.: Brouwer, Luitzen Egbertus Jan. In: Edwards (ed.) [36], pp. 399–401. Second edition in [11, vol. 4, pp. 20–57]
71. Parsons, C.: Mathematics, foundations of. In: Edwards (ed.) [36], pp. 188–213. Second edition in [11, vol. 1, pp. 700–703]
72. Posy, C.: Mathematical Intuitionism. Cambridge University Press, Cambridge (2020)
73. Richman, F.: Constructive mathematics without choice. In: Schuster, P., Berger, U., Osswald, H. (eds.) Reuniting the Antipodes. Constructive and Nonstandard Views of the Continuum, pp. 199–205. Kluwer, Dordrecht (2001)
74. Shanin, N.: Constructive Real Numbers and Constructive Function Spaces. American Mathematical Society, Providence, RI (1968)
75. Soare, R.: Computability and enumerability. In: Dalla Chiara, L., Doets, K., Mundici, D., van Benthem, J. (eds.) Logic and Scientific Methods, pp. 221–238. Kluwer, Dordrecht (1997)
76. Soerensen, M., Urzyczyn, P.: Lectures on the Curry-Howard Isomorphism. Elsevier, Amsterdam (2006)
77. Specker, E.: Nicht konstruktiv beweisbare Sätze der Analysis. J. Symb. Log. **14**, 145–158 (1949)
78. Swaen, M.: Het vinden van $\epsilon$-dekpunten. Nieuw Archief voor Wiskunde **14**(4), 275–278 (2013). Review of [99]
79. Taschner, R.: The Continuum. A Constructive Approach to Basic Concepts of Real Analysis. Vieweg, Wiesbaden (2005)
80. Troelstra, A.: Principles of Intuitionism. Lecture Notes in Mathematics, vol. 95. Springer, Berlin (1969)
81. Troelstra, A.: Choice Sequences. A Chapter of Intuitionistic Mathematics. Oxford University Press, Oxford (1977)
82. Troelstra, A., Schwichtenberg, H.: Basic Proof Theory, 2nd edn. Cambridge University Press, Cambridge (2000)
83. Troelstra, A., van Dalen, D.: Constructivism in Mathematics. An Introduction. North-Holland, Amsterdam (1988)
84. van Atten, M.: On Brouwer. Wadsworth, Belmont (2004)
85. van Atten, M.: Brouwer Meets Husserl. On the Phenomenology of Choice Sequences. Springer, Dordrecht (2007)
86. van Atten, M.: The hypothetical judgement in the history of intuitionistic logic. In: Glymour, C., Wang, W., Westerståhl, D. (eds.) Logic, Methodology, and Philosophy of Science XIII: Proceedings of the 2007 International Congress in Beijing, pp. 122–136. College Publications, London (2009)
87. van Atten, M.: Essays on Gödel's Reception of Leibniz, Husserl, and Brouwer. Springer, Dordrecht (2015)
88. van Atten, M.: The Brouwer-Kripke Schema, the Creating Subject, and infinite proofs. Indag. Math. **29**, 1565–1636 (2018)
89. van Atten, M.: Luitzen Egbertus Jan Brouwer. In: Zalta, E. (ed.) The Stanford Encyclopedia of Philosophy, Spring 2020 edn. Metaphysics Research Lab, Stanford University (2020). https://plato.stanford.edu/archives/spr2020/entries/brouwer/
90. van Atten, M.: The development of intuitionistic logic. In: Zalta, E. (ed.) The Stanford Encyclopedia of Philosophy, Summer 2022 edn. Metaphysics Research Lab, Stanford University (2022). https://plato.stanford.edu/archives/sum2022/entries/intuitionistic-logic-development/
91. van Atten, M.: Intuition, iteration, induction. Philos. Math. **32**(1), 34–81 (2024)

92. van Atten, M., van Dalen, D.: Arguments for the continuity principle. Bull. Symb. Logic **8**(3), 329–347 (2002)

93. van Atten, M., Sundholm, G.: L. E. J. Brouwer's "Unreliability of the logical principles". A new translation, with an introduction. History Philos. Logic **38**(1), 24–47 (2017)

94. van Dalen, D.: The use of Kripke's schema as a reduction principle. J. Symb. Log. **42**(2), 238–240 (1977)

95. van Dalen, D.: From Brouwerian counter examples to the Creating Subject. Stud. Logica. **62**(2), 305–314 (1999)

96. van Dalen, D.: The Dawning Revolution. *Mystic, Geometer, and Intuitionist. The Life of L E. J. Brouwer*, vol. 1. Clarendon Press, Oxford (1999)

97. van Dalen, D.: Hope and Disillusion. *Mystic, Geometer, and Intuitionist. The Life of L. E. J. Brouwer*, vol. 2. Clarendon Press, Oxford (2005)

98. van Dalen, D.: How to constructivize Brouwer's fixed point theorem. In: International Conference on Systemics. Cybernetics, and Informatics (ICSCI2009), pp. 654–656. Pentagram Research Centre, Hyderabad (2009)

99. van Dalen, D.: Intuïtionistische analyse. Een constructief denkraam. Epsilon Uitgaven, Utrecht (2010)

100. van Dalen, D.: The Selected Correspondence of L.E.J. Brouwer. Springer, London (2011)

101. van Dalen, D.: L. E. J. Brouwer—Topologist, Intuitionist, Philosopher. How Mathematics is Rooted in Life. Springer, London (2013). Second, revised edition, in one volume, of [96] and [97]

102. van Dalen, D.: Logic and Structure, 5th edn. Springer, Berlin (2013)

103. Van Heijenoort, J. (ed.): From Frege to Gödel: A Sourcebook in Mathematical Logic, 1879–1931. Harvard University Press, Cambridge MA (1967)

104. van Oosten, J.: Realizability. An Introduction to its Categorical Side. Elsevier, Amsterdam (2008)

105. van Stigt, W.: Brouwer's Intuitionism. North-Holland, Amsterdam (1990)

106. Veldman, W.: MR3186830. Mathematical Reviews, review of [99]

107. Veldman, W.: Intuitionism. An Inspiration? Jahresbericht der Deutschen Mathematiker Vereinigung **123**, 221–284 (2021). arxiv:abs/2102.01561

108. Verburgt, L., Hoppe-Kondrikova, O.: On A. Ya. Khinchin's paper "Ideas of intuitionism and the struggle for a subject matter in contemporary mathematics" (1926). A translation with introduction and commentary. Historia Mathematica **43**, 369–398 (2016)

109. de Vrijer, R., Nederpelt, R., Geuvers, J.: Selected Papers on Automath. Elsevier, Amsterdam (1994)

110. Waaldijk, F.: On the foundations of constructive mathematics—especially in relation to the theory of continuous functions. Found. Sci. **10**(3), 249–324 (2005). https://www.fwaaldijk.nl/mathematics.html

**Symbols**
#, 53
$\square_n$, 120
$\forall\alpha\exists\beta$-continuity, 122
$\forall\alpha\exists x$-continuity, 110, 112, 122
$\forall$-PEM, 118
$\Vdash$, 145, 146
$\lambda$-abstraction, 41, 143
$\mathbb{N}$, 36
$N_{20}$, 49
$\omega$, 108
$p : P$, 17
$\mathbb{Q}$, 36
$\mathbb{Q}^c$, 127
$\mathbb{Q}^{cc}$, 127
$\mathbb{R}$, 45
$\mathbf{0}$, 123
$\mathbb{Z}$, 36

**A**
Absolute value, 56
Absolutely convergent, 61
Absolutely undecidable proposition, 19
Absolutely unsolvable problems, 18, 19
Abstraction, 1, 133
AC, *see* Choice, Axiom of
AC-NN, *see* Choice, Axiom of (AC):
    Countable
AC-NN!, *see* Choice, Axiom of (AC):
    Countable Unique
AC-RN, *see* Choice, Axiom of (AC): from
    $\mathbb{R}$ to $\mathbb{N}$

AC-00, *see* Choice, Axiom of (AC):
    Countable
Addition
  of real numbers, 51
Alexandrov, P., 133
Algorithms, 1, 2, 28, 130, 131, 133–135
  and decidability, 36
Apartness
  definition of, 53
  and equality, 55
  and inverses, 55
  and multiplication, 56
  and operations, 56
  and palpable difference, 55
  properties of, 54
  strong counterexample to equivalence to
    inequality, 121
Approximate Fixed Point Theorem, *see*
    Fixed Point Theorem: constructive
Arithmetization, 40
Atomic proposition, 18
Atomic statement, 16
Automath, 145
Axiom, *see* Choice, Axiom of (AC); Creat-
    ing Subject: axioms for (CS); separa-
    tion, Axiom of

**B**
Baire space
  natural topology of, 112
  and the universal spread, 110
Bar, 114



© The Editor(s) (if applicable) and The Author(s), under exclusive license to Springer
Nature Switzerland AG 2026

D. van Dalen et al., *Intuitionistic Analysis*, Springer Undergraduate Mathematics Series,
https://doi.org/10.1007/978-3-032-16491-9

Bar Theorem, 114, 137
Barendregt, H., 11
Barycentric coordinates, 91, 99, 100
  normalized, 92, 93, 99
Barzin, M., 20
Beth, E., 145
BHK, *see* Brouwer-Heyting-Kolmogorov
    Interpretation
Bi-implication, 8
Bijection, 36
  bi-continuous, 93, 100
  from Cartesian to barycentric coordinates,
      92, 100
  from $\mathbb{N}$ to $\mathbb{N} \times \mathbb{N}$, 37
  from $\mathbb{N}$ to $\mathbb{Q}$, 37
  from $\mathbb{N}$ to $\mathbb{Z}$, 36
Bishop's mathematics
  as the common part, 136
  formalizations of, 137
  weak counterexamples in, 136
Bishop's school, 136
Bishop, E., 5, 105, 136
Bivalent logic, 9
BK-sequence, *see* Brouwer-Kripke
    sequence
BKS, *see* Brouwer-Kripke Schema
Bolzano-Weierstrass Theorem, 59
  weak counterexample to, 59
Borel, É., 1
Bounded
  Cauchy sequence, 51, 53
  monotone ascending sequence, 58
  set, 66
  weak counterexample to "bounded
      implies existence of supremum", 66
  *see also* totally bounded
Brouwer, L. E. J.
  at the origin of the Proof Interpretation, 15
  choice sequences, 106
  classical topology, 91
  collections, 108
  continuity, 82, 117
  Continuity Principle, 112
  continuum, 39, 109, 119
  Creating Subject, 120
  dissertation of, 2, 106, 108
  first and second acts of intuitionism, 105,
      106
  Fixed Point Theorem, 91
  on implication, 16
  initiates comprehensive constructivism, 2
  Kant, 39, 119
  logic, 2, 7, 14, 15, 23, 139
  logical theorem proved by, 19
  on mathematics and logic, 23
  on the nature of natural numbers, 25
  notation for ordering, 46
  rejection of bivalence, 15
  rejection of Hilbert's Dogma, 18
  Schopenhauer, 119
  species, 6, 108
  spreads, 6, 109
  supervisor of, 23
Brouwer-Heyting-Kolmogorov
    Interpretation, 15
    *see also* Proof Interpretation
Brouwer-Kripke Schema (BKS), 135
  conflicts with $\forall\alpha\exists\beta$-continuity, 122, 123
  contradicts CT, 123
  and the Creating Subject, 121
  definition of, 121
  and discontinuous functions, 125
  with MP formally implies PEM, 135
  and subspecies of $\mathbb{Q}$, 122
  and unsplittability, 126
Brouwer-Kripke sequence, 121, 125, 126,
    135
Bruijn, N. de, 145


**C**
Cantor, G., 37, 107, 110
Cantor space, 110
Cauchy sequence
  and AC-NN, 42
  boundedness of, 51
  and choice sequence, 107
  classical vs constructive, 42
  converging to a rational, 51
  conversion to a decimal expansion, 134
  definition of, 42
  distance to its elements, 57
  equivalence class of, 45
  equivalence of, 43–45
  of functions, 62
  and the inverse, 55
  not a unique representation of a real, 43
  operations on, 51
  of real numbers, 58
  role of, 43
Choice, Axiom of (AC), 40, 128, 132, 136
  Countable (AC-NN), 40–42, 128, 129
    and Cauchy sequences, 42
  countable unique (AC-NN!), 132
  from $\mathbb{R}$ to $\mathbb{N}$ (AC-RN), 128
    strong counterexample to, 128

implies PEM (Diaconescu's Theorem),
129
weak counterexample to, 40, 129
Choice function, 41, 129
Choice sequence
and Cauchy sequence, 107
and continuity, 112
definition of, 106
and discontinuity, 124
as a function, 41, 111
operations on, 107
as a path through a tree, 110
undecidability of equality to **0**, 125
Church, A., 130
Church-Turing Thesis (CT), 123, 125, 130,
133
contradicted by BKS, 123
and weak counterexample, 135
Classical logic, 7, 9, 23, 131, 141
Classical mathematics
Bishop's constructivism as part of, 136
condition for divisibility, 55
connectedness of $\mathbb{R}$, 68
continuous functions on closed intervals,
85
different but equivalent ways of obtaining
$\mathbb{R}$, 134
equivalence of least number principle and
induction, 30
Fixed Point Theorem, *see* Fixed Point Theorem: classical
formally contradicted by intuitionistic
mathematics, 118
incompatible with Brouwer's proof of the
Bar Theorem, 137
influence of logic on convergence theorems, 58
Intermediate Value Theorem, 75, 94
inverses in, 53
not contradicted by Bishop's constructivism,
136
position of logic in, 130
properties of operations on real numbers
do not always transfer, 57
splittability as a logical fact, 127
and strong extensionality, 74
uniqueness of the empty set, 63
use of AC in, 40
validity of trichotomy in, 48
Classical propositional logic, 20
Classical set theory, 30
Closed interval, 50, 51, 59, 115, 117
Compactness, 68, 94, 114, 117, 119

Comparison Test, 61, 62
Complement, 123
Completeness of the real numbers, 58
Complex numbers, 34
Computability, 28, 130
Computable function, 130
Conditional convergence, 62
Conjunction, 8, 9
Connected, 68, 127
Construction
as human activity, 2, 5, 105
mental, 15, 26, 145
method, 16, 17, 108
as a function, 42
as an operation, 129
process, 31
proof as, 15, 145
*see also* creation
Constructive Fixed Point Theorem, *see*
Fixed Point Theorem: constructive
Constructive function, 130
Constructive logic, 7, 15, 19, 23
Constructive mathematics without AC-NN,
40
Continuity
contrasted to denseness, 40
of a real function, defined, 72
uniform, 82, 84–86, 97, 113
Continuity Principle, 110, 112, 113
$\forall\alpha\exists\beta$, 122
conflicts with BKS, 122, 123
$\forall\alpha\exists\beta$
strong counterexample to, 123
$\forall\alpha\exists x$, 110, 112, 122
Continuity Theorem, 113, 128
contradicts classical mathematics, 118
uniform, 117
and unsplittability, 118
Continuum, 39, 40
of intuitive time, 39, 106
not composed out of discrete elements, 39,
109
of physical time, 39
spatial, 39
temporal, 39
is unsplittable, 118
Contraposition
everyday weak counterexample to, 151
Convergence, 58
of a sequence of functions, 62
Convergence rate, 45
Convergence theorem, 58
Convergent sequence, 58

Convergent series, 61
Convertibility of $\neg\forall x\neg$ to $\exists x$
    weak counterexample to, 19
Countable choice, axiom of, *see* Choice,
        Axiom of (AC): Countable
Countable Unique Choice, Axiom of,
        *see* Choice, Axiom of (AC): Countable
    unique
CP, *see* Continuity Principle
Creating Subject
    axioms for (CS), 120
        entail BKS, 123, 135
    and calculability, 123
    and free will, 107
    is an idealization, 17
    method to construct counterexamples,
        120
    sequence of acts of, 120
    *see also* Brouwer-Kripke Schema (BKS)
Creation, 25, 26, 106, 147
    *see also* construction
CS, *see* Creating Subject: axioms for
CT, *see* Church-Turing Thesis
Curry-Howard isomorphism, 145
Cut-off subtraction, 27


**D**
Darwin, C., 71
Decidability, 4, 30, 31, 36, 130
Decidability of equality of choice sequences
        to **0**
    weak counterexample to, 125
Decimal expandability
    strong counterexample to (in Markov anal-
        ysis), 134
    weak counterexample to, 49, 134, 135
Decimal expansion, 49, 73, 134, 135
    obtained from a Cauchy sequence, 134
    of $\pi$, 3, 4, 7, 15, 16, 63, 115
Dedekind cuts, 134
Definition
    by induction, 26
    by recursion, 26, 28, 29, 31
De Morgan's laws, 12, 13, 149, 150
Demuth, O., 134
Dense ordering without endpoints, 37
Denumerable, 36
Denumerable dense ordering without end-
        points, 37
Derivative, 86
Diaconescu, R., 129
Diaconescu's Theorem, 129, 130

Diagonal argument, 110
Differentiability
    implies continuity, 86
    uniform, 86
Differentiation
    iterated, 89
Discontinuity, 124, 125
Discontinuous function
    weak counterexample to existence of, 128
Discrete
    not composed out of continua, 39
    set, 4
    *see also* twoity; continuum
Disjunction, 8, 9, 13
Distance, 28, 67
Divergent, 61
Domain
    and AC-NN, 42, 129
    of discourse, 10
    and fixed points, 81
    in Kripke models, 146, 147
    of quantification, 10, 17
    of species, 108
    and variables, 28
    and well-definedness of a function, 29
Double negation, 11, 13, 19, 21
    weak counterexample to, 4
Du Bois Reymond, E., 18


**E**
Effective relations, 31
Effectivity, 36
Elimination rule, 140
Embedding, 35
Empty, 63
Equality
    and apartness relation, 55
    of choice sequences, 124
    decidable, 4, 33, 114, 125, 129
    of natural numbers, 31, 129
    of real numbers, 51
        is undecidable, 48, 64, 128
    *see also* discrete set
Equivalence, 8, 9, 35
    natural numbers not constructed via, 129
Errera, A., 20
*Ex Contradictione*, 11
*Ex Falso*, 11
    as counterfactual, 17
*Ex Falso*, 17
Excluded Middle, *see* Principle of the Excluded
        Middle (PEM)

Existence, 17
Existential quantifier, 8
Exponentiation, 27
Extensionality, 74

**F**
Factorial, 27
Falsum, 8, 10, 11, 16, 18
Fan, 110
    binary, 110, 111
    ternary, 115
Fan Theorem, 113, 114, 119, 136
Finite, 65
Finite sequences, 112
Fixed point, 81
Fixed Point Theorem
    classical, 91
    constructive (or approximate), 91
        one-dimensional, 97
        two-dimensional, 99
    weak counterexample to, 81, 94
Forcing, 145, 146
"free for", 142
Function
    computable, 130
    differentiable, 86
        uniformly, 86, 89
    method turned into, 41, 143
    monotone, 78
    nowhere differentiable, 133
    vs operation, 129
    well-definedness of, 29
Functional, 111
Fundamental Theorem of Algebra, 5

**G**
Gentzen, G., 139
Glivenko, V., 19, 20, 133
    Theorem, 19
Gödel, K., 130, 133
Goldbach Conjecture, 14, 15, 42
Goodman, N., 129
Goodstein, R., 134
Greatest lower bound, *see* infimum
*Great Soviet Encyclopedia*, 136

**H**
Heine-Borel Theorem, 83, 117, 119
    weak counterexample to, 83
Henkin, L., 131
Heyting, A., 15, 16, 85, 133, 139

Hilbert, D., 14, 18, 133
Hilbert's Dogma, 18, 19
Husserl, E., 39

**I**
Ideal mathematician, *see* Creating Subject
Ideal subject, *see* Creating Subject
Ignorabimus, 18
Implication
    classical, 9, 10, 13
    is the key concept of logic, 16
    notation, 8
    Proof Interpretation of, 16
    and universality, 17
Index-finite, 65
    weak counterexample to "index-finite
        implies finite", 66
Indirect reasoning, 21, 47
Induction, *see* Principle of Complete Induc-
    tion
Induction hypothesis, 29
Inductive definition, *see* definition: by induc-
    tion
Inequality
    strong counterexample to equivalence to
        apartness, 121
Infimum (inf), 66, 67, 119
Infinite path, 110
Infinity, 10, 14, 21, 34
Inhabited, 64
Integers
    basis for $\mathbb{Q}$, 35
    construction from $\mathbb{N}$, 35
    decidable equality of, 33
    as equivalence classes, 35
    neat sets of, 64
    set of, 36
Intended interpretation, 15
Interior, 118
Intermediate Value Theorem
    approximate, 80, 94, 95
    classical, 75, 94
    extension of the hypothesis, 104
    weak counterexample to, 76, 80, 85
Introduction rule, 140
Intuition
    basic, 108, 109, 119, 120
    of the passage of time, 25
    *see also* continuum; time: intuitive; two-
        ity
Intuitionistic logic, 7
    no definitive formalization, 141

is open-ended, 141
Intuitionistic mathematics
   Bishop's constructivism as part of, 136
   formally contradicts classical mathematics, 118
   logic as a distraction from, 23
   is open-ended, 141
Intuitionistic propositional logic, 20
Intuitive time, *see* time: intuitive
Inverse
   and apartness, 55
   of a rational, 21
   of a real, 55
Iteration, 26, 108

**K**

Kant, I., 39, 119
Khinchin, A., 133
Kleene, S., 130
KLST, *see* Kreisel-Lacombe-Shoenfield-
   Tseitin Theorem
Kolmogorov, A., 15, 20, 133
König's lemma, 114
Korteweg, D., 23
Kreisel, G., 11, 120
Kreisel-Lacombe-Shoenfield-Tseitin Theo-
   rem (KLST), 131
Kripke model, 147
   and weak counterexamples, 150
Kripke semantics, 139, 145, 149
Kripke, S., 120, 145
Kripke's Schema (KS), *see* Brouwer-Kripke
   Schema
Kronecker, L., 1
KS, *see* Brouwer-Kripke Schema (BKS)

**L**

Least Number Principle, 22, 30
   weak counterexample to, 30
Least upper bound. *see* supremum
Leibniz, G.W., 71
Limit, 36, 58
Limit point, 59
Linnaeus, C., 71
Located, 65, 68
Logical connectives, 8
Logical operator, 8–10, 17, 18

**M**

Markov, Jr., A., 5, 105, 133, 134
   on choice sequences, 136

Markov's Principle (MP), 132, 135
   with BKS formally implies PEM, 135
   with recursive function, 132
Markov, Sr., A., 133
Martin-Löf, P., 145
Mathematical subject, *see* Creating Subject
Maximum, 27, 56, 85
   and weak counterexample, 57
Mental acts, 106, 120
Method, *see* construction: method
Mind, 25
Minimum, 27, 56, 85
Modulus function, 86
Molk, J., 1
Monus, 27
Mostowski, A., 134
MP, *see* Markov's Principle
Multiplication
   and apartness, 56
   of natural numbers, 26
   of real numbers, 51
   and weak counterexample, 57
Myhill, J., 120, 129

**N**

Natural deduction, 139
Natural numbers
   and AC-NN, 40, 42
   basis for $\mathbb{Z}$ and $\mathbb{Q}$, 35
   canonical form, 26, 129
   construction of, 25, 129, 132
   decidable equality of, 33, 114
   finite sequences encoded as, 112
   natural order, 25
   neat sets of, 64
   notation, 26
   sequence of, 106, 120, 132
   set of, 36
   universal statements about, 17
   and well-ordering, 30
Negation
   classical, 10, 13
   Proof Interpretation of, 21
Negative subspecies, 126, 127
Net, 68
Non-empty, 30, 40, 63, 64, 68
   weak counterexample to "non-empty implies
      inhabited", 63

**O**

Omniscience principle, 18
One-one correspondence, 36

Open interval, 50
Operation
  vs function, 129
Ordered pair, 41, 91
Ordinals
  intuitionistic, 108

**P**
Pair-forming operator, 41, 144
Palpable difference, 55
Partial ordering, 147
Partial sum, 61
Passage of time, 25
PEM, *see* Principle of the Excluded Middle
Phenomenology, 39
Platonists, 5
Poincaré, H., 1
Positive connectedness, 68
Possible worlds, 151
Power notation, 26
Power series, 62
Power species, 121, 122
PRA, *see* Primitive Recursive Arithmetic
Predecessor, 27
Predicative, 108
Primitive Recursive Arithmetic (PRA), 31
Principle of Complete Induction, 28–31
  and recursive definition, 29
Principle of Constructive Choice, 132
Principle of the Excluded Middle (PEM)
  Brouwer's early construal of, 18
  classical tautology, 13
  constructively incorrect, 14, 19
  and decidability, 30
  and equality of natural numbers, 31
  first intuitionistic target, 14
  and the Fixed Point Theorem, 94
  formally implied by BKS and MP, 135
  and Hilbert's Dogma, 18
  Hilbert on, 14
  implied by AC (Diaconescu's Theorem),
    129
  and indirect proof, 21
  and the Proof Interpretation, 18
  and splittability, 127
  and three-valued logic, 19
  universally quantified predicate-logical
    version, 118
    strong counterexample to, 118
  unreliability of, 23
  weak counterexample to, 15, 42
  and weak negation, 20

Principle of the Smallest Counterexample,
    30
Product, 26, 28
Projection operator, 41, 144
Proof
  abstract, 15
  as construction, 15
  by contradiction, 34
  by contradiction, 21, 22, 141
  by induction, 26
  as opposed to truth value, 20
  on paper, 15
  in the Proof Interpretation, 17
Proof Interpretation, 15, 17, 42, 129, 139,
    144, 145
  and AC-NN, 40
Propositional connectives, 9

**Q**
Quantifiers, 10, 142
  *see also* domain

**R**
Range, 29
Ratio Test, 62
Rational numbers
  construction from $\mathbb{Z}$, 35
  construction from integers, 35
  decidable equality of, 33
  as elements of Cauchy sequences, 51
  embedding in the real numbers, 43
  encoding in natural numbers, 114
  neat sets of, 64
  set of, 36
Real closed, 91
Real numbers, 33, 43, 129
  completeness of, 58
  as equivalence classes of Cauchy sequences,
    45, 51
  generally not rational, 43
  not denumerable, 110
  opposite of, 51
  strongly irrational, 127
Realists, 5
Recursion, 26, 28, 29
Recursion theory, 125, 130, 133, 134
Recursive analysis, 131, 134
Recursive definition, *see* definition: by recur-
    sion
Recursive functions, 125
Recursively enumerable, 123
Recursive mathematics

Bishop's constructivism as part of, 136
Reductio Ad Absurdum (RAA), 4, 94, 141
  *see also* proof: by contradiction
Representative, 129
Riemann Hypothesis, 19, 63, 66, 122, 131
Rolle's Theorem, 86

**S**
Schopenhauer, A., 119
Scott, D., 128
Search, 5
Separation, Axiom of (Aussonderung), 108
Sequence
  vs series, 61
Series, 61
  absolutely convergent, 61
  convergent, 61
  divergent, 61
  vs sequence, 61
Set theory, 23, 36
Shanin, N., 134
Simplex, simplices, 99
Singleton, 122
Size, sameness of, 36
Species
  analogy to separation, 108
  and BKS, 122
  Brouwer's introduction of, 108
  compact, 117
  defined, 107
  domain of, 108
  elements of, 107
  power species, 121
  predicativity of, 108
  singleton, 122
  spread as a special case, 109
  uncountable, 109
Specker, E., 134
Sperner labeling, 96, 99, 102, 103
Sperner's Lemma, 91, 95–99, 101
Splittability, 118, 127
  a logical fact in classical mathematics, 127
Spread, 109, 110
  elements of, 110
  as a species of choice sequences, 109
Spread tree, 110
Stable, 55
Strong counterexample
  to $\forall\alpha\exists\beta$-continuity, 123
  to $\forall$-PEM, 118
  to AC-RN, 128
  compared with weak counterexample,
    118

to decimal expandability (in Markov analysis), 134
  to equivalence of inequality and apartness, 121
Strong extensionality, 74
  of functions
    weak counterexample to, 74
Strong irrationality, 127
Strongly extensional discontinuous functions
  weak counterexample to existence of, 126
Subject, *see* Creating Subject
Successor, 25, 26
Sum, 26
Summation, 27
Supremum (sup), 66, 119

**T**
Third (truth value), 19, 20
Three-valued, 20
Threeity, 106
Time
  intrinsic to consciousness, 39
  intuitive, 25, 39, 106
  passage of, 25, 39, 106
  physical, 39
  scientific, 39
Topology
  and choice sequences, 110
  classical, 68, 91
  Euclidean, 83
  relative, 83
  of a spread tree, 110
  of a tree, 109
  *see also* Baire space; Cantor space
Totally bounded, 68, 69, 85
Tree, 109
  natural topology of, 109
Triangle inequality, 57, 88
  reverse, 58
Triangulation, 96, 97, 99
Trichotomy, 48
  weak counterexample to, 4, 49
Troelstra, A., 120
True, 11
Truth
  as correspondence, 15
  as provability, 15
Truth table, 9–11, 13, 15
Truth value, 9, 16, 19, 20
Turing machine, 36, 123, 130, 131
  as an ideal computer, 17

Turing, A., 130
Twoity, two-ity
  earlier part and later part, 106
  empty, 106, 107
  priority over "one thing", 106
  result of moment of life falling apart, 105
  of two previous systems, 106
  vs unity, 106, 107
Type theory, 145
Typed lambda-calculus, 145


**U**
Unbounded
  iteration, 26
  search, 5
Uniform differentiability, 86
Unity
  empty, 107
  vs two-ity, 106, 107
Universality, 17
Universe, 108
Unreliability
  of $\neg\forall x \neg A(x) \to \exists x A(x)$, 151
  of PEM, 19, 23
Unsplittability, 118
  of the continuum, 118
    contradicts classical mathematics, 118
  of negative, dense subspecies of $\mathbb{R}$, 126
Urysohn, P., 133
Uspensky, Y., 134


**V**
Van Dalen, D., 120
Variables, 28
  binding of, 142, 144


**W**
Weak counterexample
  to AC, 40, 129
  in Bishop's mathematics, 136
  to Bolzano-Weierstrass Theorem, 59
  to "bounded implies existence of supremum", 66
  and Church-Turing Thesis, 135
  compared with strong counterexample, 118
  to convertibility of $\neg\forall x \neg$ to $\exists x$, 19
  to decidability of equality of choice sequences to $\mathbf{0}$, 125
  to decimal expandability, 49, 134, 135
  to double negation, 4
  everyday counterexample to contraposition, 151
  to existence of discontinuous function, 128
  to existence of strongly extensional discontinuous functions, 126
  to Fixed Point Theorem, 81, 94
  to Heine-Borel Theorem, 83
  to "index-finite implies finite", 66
  to the Intermediate Value Theorem, 76, 80, 85
  introduction of, 3
  and Kripke models, 150
  to Least Number Principle, 30
  and maximum, 57
  and multiplication, 57
  to "non-empty implies inhabited", 63
  to PEM, 15, 42
  to strong extensionality of functions, 74
  to trichotomy of reals, 4, 49
Weak induction, 31
Weak negation, 20
Well-ordering
  classical, 30
  constructive or intuitionistic, 30, 108
  of the natural numbers, 30, 108
Witness, 41, 42


**Z**
Zaslavsky, I., 134
Zermelo, E., 40
Zero sequence, $\mathbf{0}$, 123